Praise for *Good to Go*

"An intelligent and entertaining tour of fitness research for anyone who exercises, with clear advice on what actually works to aid recovery." —Julia Belluz, *Vox*

"When a work like this is written by one of the best science writers in the world, there is much to learn not only from the way [Christie Aschwanden] combines narrative with a clear synthesis of what the scientific evidence actually supports, but also about how the world of sports recovery can teach us something more fundamental about human nature."
—Jonathan Wai, *Psychology Today*

"A useful introduction to how scientific research works—and why, in sports science, it often doesn't. Such insights make *Good to Go* appealing to more than just gym rats and weekend warriors. It's for anyone who wonders how scientific studies happen, and how they influence the claims on products found in grocery stores and athletic stores alike."
—Bethany Brookshire, *ScienceNews*

"Absorbing. . . . Aschwanden separates the facts from the hype in the realm of athletic recovery, dispensing welcome doses of common sense." —David Takami, *Seattle Times*

"An amusing and exhaustive takedown of the recovery products and trends that fitness enthusiasts have transformed into a multibillion-dollar industry."
—Andrea Gawrylewski, *Scientific American*

"Aschwanden . . . never loses sight of the gap between anecdote and randomized trial. Ultimately, she pins her conclusions on the best studies in each field, as well as her interviews with highly-regarded researchers."—Amby Burfoot, *Runner's World*

"[Aschwanden] wades through the Olympic-size pool of scientific research on exercise recovery. Readers are rewarded with new knowledge but left rethinking their approaches to dealing with concerns like hydration, inflammation, injury prevention and refueling."   —Andrew Yee, *Cyclocross Magazine*

"Aschwanden engagingly zooms in on the neglected topic of exercise recovery. . . . Slicing through all the fads and hoopla, *Good to Go* reinforces the absolute necessity of listening to and trusting your body."   —Tony Miksanek, *Booklist*

"Christie Aschwanden is the real deal, an inspiring champion of good science. What makes *Good to Go* so delightful is how much fun it is to read her equally persuasive debunking of pseudoscience. A rollicking read for both weekend warriors and anyone who's ever wondering what the heck is up with infrared pajamas, anyway."

—Seth Mnookin, best-selling author of *The Panic Virus*

GOOD
TO GO

GOOD TO GO

What the Athlete in All of Us Can Learn from
the Strange Science of Recovery

Christie Aschwanden

W. W. NORTON & COMPANY

Independent Publishers Since 1923

For information about permission to reproduce selections from this book, write to
Permissions, W. W. Norton & Company, Inc., 500 Fifth Avenue, New York, NY 10110

For information about special discounts for bulk purchases, please contact
W. W. Norton Special Sales at specialsales@wwnorton.com or 800-233-4830

Manufacturing by LSC Harrisonburg
Book design by Chris Welch
Production manager: Anna Oler

Library of Congress Cataloging-in-Publication Data

Names: Aschwanden, Christie, author.
Title: Good to go : what the athlete in all of us can learn from the strange science of
recovery / Christie Aschwanden.
Description: First edition. | New York : W. W. Norton & Company, [2019] | Includes
bibliographical references and index.
Identifiers: LCCN 2018050104 | ISBN 9780393254334 (hardcover)
Subjects: LCSH: Sports—Physiological aspects. | Sports medicine—Popular works. |
Sports injuries—Popular works. | Athletes—Health and hygiene. | Stress (Physiology)
Classification: LCC RC1235 .A78 2019 | DDC 617.1/027—dc23
LC record available at https://lccn.loc.gov/2018050104

ISBN 978-0-393-35771-4 pbk.

W. W. Norton & Company, Inc., 500 Fifth Avenue, New York, N.Y. 10110
www.wwnorton.com

W. W. Norton & Company Ltd., 15 Carlisle Street, London W1D 3BS

For Dave

CONTENTS

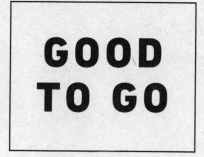

GOOD
TO GO

Introduction

was feeling pretty spent on the Saturday morning in late summer when I first visited Denver Sports Recovery. I'd just finished a 5K race, which I'd sprinted hard enough to win. My first-place finish was less impressive than it sounds—the event was billed as a beer run, with the emphasis in that order. The start/finish line was in front of a microbrewery that served free pints afterward. A guy in his twenties with the scrawny build of a distance runner and a cross-country shoe icon tattooed on his leg was the only other person I saw bothering to warm up for the race. I didn't care if we were the only two pushing the pace. I was on a summer-long mission to run a fast 5K, so I was running against myself and the clock. I finished with all I had, and afterward, my middle-aged legs were pounding and aching. Tempting as it was, I decided to skip the beer. I lacked a designated driver and it seemed like a good time to head over to Denver Sports Recovery and to try out their motto: "Recover like a pro!"

Tucked into an upscale district near downtown, Denver Sports Recovery, with its understated gray exterior, could easily be mistaken from the street for a yoga studio or CrossFit gym (or box, as they're called). I stepped inside to discover a playground for rejuvenation, equipped with every gadget and tool athletes and exercisers might need to practice the art of recovery. A rack along one wall held an array of rollers, balls, and other devices meant to squish, press, and massage achy muscles, and a spacious floor area offered room to stretch and experiment with a collection of muscle massagers. Therapy tables and lounge chairs were lined up in rows leading to a set of hot and cold tubs. An upstairs alcove held a hyperbaric (pressurized oxygen) chamber and there was a sauna and a cryotherapy tank, which uses liquid nitrogen to cool sore muscles. In an adjacent lounge painted to look like a forest, a companion business offered a chance to lie in the comfort of a recliner while receiving an IV filled with vitamins and anti-inflammatories purported to aid recuperation.

In the main recovery space, framed and autographed jerseys adorned the walls—one from former Colorado Rockies pitcher Matt Belisle, a Broncos jersey from Wes Welker, and a Jason Richardson jersey from his time with the 76ers, among others. A poster near the entry listed the available recovery tools: sauna, hot/cold contrast, compression therapy, vibration therapy, Bio-Mat, soft tissue tools, E-Stim, Marc Pro, and infrared therapy.

Before I could get lost in the sea of unfamiliar choices, a cheerful, athletic, ponytailed woman in her twenties welcomed me and said she'd be my "recovery assistant." She handed me a clipboard, a pen, and a two-page form that was a mashup of the kind of checklist you get at a doctor's office and the enrollment form you get at a gym. What sports did I do? Did I have any injuries? Would I be wanting any physical therapy or massage today? After looking over my answers, my assistant asked how much time I had, then laid out a plan for me. I would spend well over an hour doing a bunch of stuff that would help me recover from an event that, if you throw in my warm-up and cool-down, had lasted maybe 45 minutes.

My visit, she said, would involve four recovery techniques, or "modalities" as they're known in recovery speak—soft tissue work, electrical stimulation, compression, and vibration. (Next time, she said, I could try some of the others, like the contrast baths or cryotherapy.) To begin, she gave me what looked like two roller skate wheels, about a foot apart and connected with a padded axel, and instructed me to roll my foot back and forth on the axel. It felt nice—like a self-administered foot rub—and reminded me of the wooden foot rollers sold in drug stores and those catalogs found in the seat-back pockets of airplanes. After I'd given my feet a good rubdown, she moved me to a larger version of the roller, which was wider and affixed on larger wheels. Pressing my calves onto the padded cylinder, I rolled my leg back and forth, from my ankles to my knees. Having just run a pretty hard race, I felt soothed to rub my muscles like this.

While I rolled around on the floor contorted like a dog scratching itself in the dirt, my helper explained that the purpose of this rubbing and mashing was to release tension in the fascia around my muscles. Fascia is connective tissue that surrounds the muscles, and the idea behind rolling—whether with one of the fancy rollers at DSR or with a less exotic foam roller that you'd find at a physical therapist's office or yoga studio—is that it increases blood flow to the fascia and works out "adhesions" that might form. That's the theory, anyway, and the scientific-sounding explanation gave me license to pretend it was perfectly normal to come here to vigorously rub my body with exotic toys. I couldn't say whether I really had adhesions in my fascia or whether the rolling worked them out. But it did feel pretty nice on my sore muscles.

From the big roller, we moved to something smaller—a padded sphere about the size of a tennis ball, but not as soft. "You put this under here," my helper demonstrated, sticking the ball under her butt and then rolling back and forth on it as in some kind of dance move. The idea, she said, was to massage my piriformis—a muscle deep in my butt. The other rolling had felt good, but this motion hurt. It felt like I was sitting on a very hard tennis ball and jabbing it into my butt, which is pretty much exactly what I was doing. When I told my helper that this was painful, her face brightened. Oh good, she said—you've found a sore spot. Personally, I wasn't so excited. I'd come here to relieve the pains I knew about, not find new places to hurt. I hadn't even remembered that I had a piriformis until I mashed the ball into it, so when she wasn't looking, I pulled the ball out from under me.

Unlike my piriformis, my hamstrings are an ongoing source of agony, so when my recovery assistant suggested that I try electrical stimulation on them, I was game. While I lay stomach-down on a massage table, she attached electrodes to my hamstrings with sticky patches smeared with a cold gel. The electrodes connected to an e-stim machine that sent electrical pulses to my muscles. For the next 25 minutes, the machine took over my hamstrings, and without any thought or effort from me, my leg muscles contracted and relaxed in a steady rhythm. I kept expecting my leg to involuntarily kick up from the table, but that never happened. As surreal as it was to feel my leg muscles firing on the machine's command, it wasn't exactly unpleasant. After a while, I had so completely ceded control to the machine that I had stopped feeling its twitches, and by the end of my session I was on the verge of falling asleep.

I moved to the pneumatic compression boots, which are like individual sleeping bags for your legs that envelop the feet and zip up to the waist. At the turn of a switch, the bags inflated to squeeze my muscles, and presumably increase the circulation in my legs, which my aide said would reduce inflammation and lactic acid. The squeezy pants had numerous available settings, but I tried the one that pulsed in a predictable sequence. The squeezing began at my toes and feet, then moved up to my calves, knees, then upper legs, as if squeezing toothpaste from a tube. Once the entire leg was compressed for a few minutes, the pouches began to gently deflate, starting with my feet and ending at my thighs. Then the cycle began again. It felt like a very methodical massage, which is to say, it felt great.

Before ending my visit, I tried a strange device I'd been eyeing since I arrived. "The Swisswing is everyone's favorite," my recovery assistant told me as she sat me down in a chair next to the device. It looked like a giant trash can set on its side and covered with a yoga mat. The device was attached to a giant articulating arm that allowed you to adjust its angle. Sitting in the chair, I lifted my feet on top of the cylinder. At the flip of a switch, the whole cylinder shook at a fast, mesmerizing pace. The vibrations felt soothing, even hypnotic. I could almost see the soreness being shaken out of my muscles. After about five minutes on my feet, I scooched my legs further up the device so my calves were centered on the top of the cylinder. Five minutes later, I was standing next to the Swisswing, nearly straddling it, getting some good vibrations on one hamstring at a time.

I was jutting my butt onto this self-serve massage barrel when a middle-aged guy who looked like he'd been working off a small beer belly approached to ask how I liked it. This wasn't as creepy as it sounds. He was training for a marathon, he said, and had recently started coming to DSR because his long runs were beating him up. He liked the Swisswing too, and I realized that he wasn't just being friendly. He was waiting for me to finish so he could have his turn.

———

How did I end up in a fancy gym, jiggling my butt on a giant vibrator? When I was a serious athlete in the 1990s and 2000s, recovery was a noun—a state of being you hoped to attain through all the things you *weren't* doing, like training, standing around on your feet, staying out late socializing, or getting caught up in stressful activities. Recovery meant resting, and the only thing you *did* was sleep and lie back with your feet up and your nose in a book. Today, recovery has become a verb. It's something that athletes—pros and weekend warriors alike—do with almost as much gusto and drive as their training. Recovery even has its own gear now. The first time I heard someone say she needed to "go *do* my recovery," I cringed. By the tenth time, I'd come to understand that recovery is no longer a waiting period between workouts. Instead, it's become an active extension of training itself.

No matter the sport, top athletes don't just work hard, they recover hard too. Nowhere is this more apparent than on social media, where the pros regularly post images of themselves taking part in various recovery rituals (and showcasing their sponsors' recovery products). Leading up to the 2016 Olympics in Rio, there was gymnast Simone Biles showing off her NormaTec pneumatic compression boots. The most decorated Olympic athlete of all time, swimmer Michael Phelps, showed up at the pool with circular purple bruises all over his shoulders and back. His giant hickeys, about the circumference of soda cans, were the result of a practice called "cupping," in which glass cups are placed on the skin with suction that draws the skin up, bursting capillaries. Phelps swears the process aids recovery and reduces muscle soreness.

Covering your body with hickeys might sound ridiculous, but it's nothing compared to a recovery strategy that went viral when former NBA star Amar'e Stoudemire posted a selfie of himself bathing in a tub filled with red wine.[1] ("Recovery Day! Red Wine Bath!!" the caption read.) The internet erupted, and stories about the NBA standout's habit appeared in *Sports Illustrated*, *Bleacher Report*, *Deadspin*, and *New York Magazine*, to name just a few. Stoudemire was playing for the New York Knicks at the time, and ESPN sent writer Sam Alipour to interview him as the star center indulged in so-called vinotherapy at an unnamed New York City spa.[2] With UB40's "Red, Red Wine" playing in the background, Alipour, dressed in a shirt and tie, sat on the edge of the polished steel tub and questioned Stoudemire. "Let's start with the elephant in the room. Why are you bathing in red wine?" Stoudemire laughed. "Well, I feel it's great for recovery. They say drink a glass a day, keep the doctor away . . . so I take it upon myself to really submerge myself into some red wine." When Alipour asked him what it feels like to bathe in wine, Stoudemire invited him into the tub. Sliding into the tub next to the lanky NBA player, Alipour remarked, "You know what it feels like? It feels like money."[3]

Alipour was on to something. Although Stoudemire declined to name the location where he snapped his wine-soak selfie, the photo created a spike in interest regarding vinotherapy (Google searches for the term rose about 90 percent), and spas around the country took the opportunity to promote their wine-related therapies.

"The buzzword is recovery," says chiropractor Ryan Tuch-scherer, who runs a series of cryotherapy clinics around the Denver region that offer anyone willing to pay the chance to get "treated like the pros." When a star athlete comes into his clinic, Tuchscherer says, "I can take a video and put it out on Instagram and someone in Florida or Minnesota can find it and think, how can I get that? When they see the pros on social media, they see a trusted brand. It spreads like wildfire."

The numbers show a booming market for products and services claiming to help recovery. "This is not a little flash in the pan," says sports industry analyst Matt Powell of the NPD Group, who estimates the growing sector is worth somewhere in the hundreds of millions of dollars. The market includes goods and services ranging from drinks, bars, and protein shakes to compression clothing, foam rollers, ice packs, cryotherapy, massage, laser therapy, electrical muscle stimulators, saunas, float tanks, meditation videos, and sleep trackers.

The search for an extra recovery edge has spawned an arms race, with athletes looking for ways to bounce back faster and entrepreneurs rushing in with products and promises, not just for the pros, but for recreational athletes and people exercising for fitness too. Today there is a product or service marketed to address every possible aspect of postexercise recovery. Tired? Try a recovery drink to restore your energy. Sore? Soothe achy muscles with your choice of massagers, compressors, supplements, or cold treatments. Run-down? Try a massage or meditation app. Having trouble sleeping? There's a tracker and app for that too.

If all these options seem hard to navigate, services like Denver Sports Recovery are standing by ready to help. It's among a growing number of businesses that have opened to offer high school athletes, serious amateurs, weekend warriors, and master blasters a chance to get in on the fancy recovery tools their idols keep shilling on Instagram, all under the guidance of a professional trainer. The Denver/Boulder area alone has at least four of these centers, which have also popped up in New York City, California's Bay Area, Dallas, Phoenix, Chicago, Washington, D.C., and beyond.

But what exactly is this highly prized thing called recovery they're selling? The explosion of recovery products and services can seem ridiculous, because in its most basic form—a return to readiness following an intense workout or competition—everyone intuitively knows what recovery is and how to achieve it. And yet, we've somehow managed to make every aspect of it—nutrition, relaxation, and sleep—vastly more complicated, expensive, and time-consuming than it was before.

I've been an athlete since joining my high school cross-country team at age thirteen. In addition to running, I've also competed seriously in cycling and Nordic skiing, and I've dabbled in swimming and climbing too. As I've aged, I've found that it takes me longer to rebound from one workout to another. I'm a skeptic by nature, but as I've found myself needing more recovery time, those ads for recovery tools have begun to seem much more intriguing. I wanted to know: Do any of these products actually work?

Whether it's a magic drink or a space-age gadget, we're suckers for a good sales pitch, especially one that promises that it can make us better than we were before. Even as we know that a protein supplement or special compression sleeve is unlikely to prove life-changing, we hold out some glimmer of hope that the right nutrient or tool could fix what's holding us back and unlock our hidden potential. We're all seeking the secret, especially if it can be boiled down to one quick tip, or better yet, something to buy.

But the question remains: Does any of this stuff work? Has the commodification of recovery made us better off? Do these gizmos and rituals help us recover better, or could they instead be contributing to our difficulties relaxing? Most of these popular approaches to recovery come with scientific-sounding explanations, but does the evidence bear them out? How much effort should we put toward recovery? When recovery becomes a new chore, does it steal time away from real restoration? What does it really take to achieve optimal recovery and get the most out of the physical work we do?

With questions like these in mind, I set out to examine the complex physiology that determines how our bodies recover and adapt to exercise, and find out what it takes to master the fundamentals—and science—of recovery.

1

Just-So Science

The Garfield Grumble is the stupidest race I've ever done. The first year I ran it, I had no idea what I was getting myself into. I'd just moved to western Colorado, where jagged peaks intersect with the high desert and deep red rock canyons, and I'd seen a notice in the local newspaper that said simply, "5-mile trail run." It sounded like a fun way to explore my new landscape. In retrospect, I should have paid closer attention to the name. I'd later find out that the event is also known as the "summit and plummet," which is a pretty accurate description. The race begins in a dusty parking lot and then immediately rises up Mt. Garfield, ascending 2,000 feet in two miles. It's the kind of climb where using your hands helps. The course scrambles up sandstone boulders and at one point traverses a tiny plateau, where I spotted a small group of wild horses grazing. After hitting the summit, the trail drops through a hidden slot in the far ridge line and cuts across a steep, exposed face composed mostly of soft shale. Seen from afar, the slope looks impossibly steep and impassable. But descend it does. For approximately three quad-busting miles, the trail traverses and switchbacks down to the finish, which is marked by a rectangular fire pit that runners must leap over to cross the line.

My first time at the Grumble, I didn't know what lay ahead, and I went out so hard that I built a gap big enough that even with an unscheduled stop to regain my breath and my bearings, I was still the first woman to cross the finish line, which earned me a beautiful framed photograph of Mt. Garfield. (A friend later told me that a previous iteration of the event had awarded a used car to the winner. Perhaps the car also ran the course, because it was so beat up that it went home with the fourth- or fifth-place finisher, after one runner after another declined to accept the prize.) What I learned at that first Grumble is that the Mesa Monument Striders Running Club that put on the race was a friendly, fun-loving, beer-drinking group after my own heart and that running up and down Mt. Garfield is like putting a jackhammer to your legs. The day after that first race, my quads were so sore I could barely climb out of bed. For several days afterward, it hurt to even think about walking.

I was suffering from one of the most common (and painful) aftereffects of a hard exercise session—delayed-onset muscle soreness, or DOMS. The pain of DOMS typically peaks 24 to 72 hours after the exercise (hence the name), and it's more likely to occur after exercise that emphasizes eccentric contraction where the muscle is lengthening, rather than shortening (think the phase in an arm curl where you're lowering a barbell, versus raising it). It wasn't surprising that the plummet down Mt. Garfield had given me a bad case of DOMS. When scientists want to study DOMS, they typically have volunteers run downhill or jump down from a height multiple times— movements that force the leg muscles to contract while they're in a lengthening phase. These opposing actions yank at the muscles, causing microscopic tears in the muscle fibers, hence the pain.

In response to this damage, your body sends in a cleanup team to clear out the damaged tissues and rebuild the muscles, making them stronger and more resilient in the process. This repair response may at least partially explain the "repeat bout effect," which makes a second round of damaging exercise less DOMS-inducing, because, in response to the first bout, your muscle has adapted to become stronger. I've done the event multiple times since that first Grumble (a smarter person would have decided that once was enough), and I now take care to do a few fast downhill runs before the race. That strategy, plus braking a little less on the steeps, has prevented me from ever feeling quite so crushed again after the dumb race. I still get some DOMS and plenty of general postrace fatigue, but I can get out of bed the next morning without moaning, *It wasn't worth it!*

After the Grumble, runners assemble around the finish area, swapping tales about their run and the mishaps they endured. Just about every year, someone shows up bloody from a fall or grumpy about having been forced to bushwhack through sagebrush after missing the trail down from the summit. I have been that person multiple times. But there's one thing that no one seems to grumble about, and that's the cold beer at the finish. It tastes so refreshing on a hot day. "Cheers, we earned it!" we joke as we lift the cool cans to our mouths. It's not just an indulgence—we need those postexercise carbs and fluids, right?

A few years back, as I gulped my cold finish-line brew, I began to wonder: was beer really such a great recovery drink? The importance of replenishing fluids and carbohydrates had been drilled into me by a long string of magazine articles and sports drink ads, and there was no denying that beer could deliver both of those things. But the alcohol in this tasty beverage seemed potentially problematic. Could it be partly to blame for my post-Grumble DOMS and that utterly spent feeling the day after the run?

Beer is a little like coffee—it's mind-altering, pleasurable, and potentially dangerous at high doses, so we can't help suspecting that a little must be bad for us too, even as we hope that it's not. But millions of us routinely drink beer after a game or a workout. My mountain-biking buddies and I make a habit of hitting the brewery or our beer cooler after a ride. I doubt many of us expect that it's *good* for us, but a tiny part of me feared that it might slow or otherwise hurt how I recover from my favorite physical activities. At the same time, I really wanted to believe that beer might be a perfectly fine postexercise beverage, assuming you didn't overdo it, of course.

I'm not crazy to hope that beer could be a suitable recovery drink—it's got carbs and some minerals, after all, and those are ingredients we've been told are essential after exercise. A Canadian company developed a "recovery beer" called Lean Machine, and German beer maker Krombacher supplied about 3,500 liters of their nonalcoholic beer to the German Olympic team during the 2018 Pyeongchang Games.[1] The German Olympic ski team's doctor, Johannes Scherr, told the *New York Times* that nearly all of his athletes drink nonalcoholic beer. Krombacher isn't the only beer targeted at athletes; the nonalcoholic beer Erdinger Alkoholfrei is also marketed as a recovery drink. According to advertisements, Erdinger's recovery beer "replenishes the body with essential vitamins including B9 and B12, which help reduce fatigue, promote energy-yielding metabolism and support the immune system." Erdinger's sport beer has less than 0.5 percent alcohol by volume and 125 calories per serving, which is probably not all that much different from what's in the light beers we were quaffing at the Grumble finish line.

Marketing aside, I wondered if there was any science to show that beer could help or hurt recovery. So I dug into the scientific literature. Turns out, there wasn't much research to answer my question. I found a few studies looking at athletic performance while under the influence of alcohol (apparently brandy was part of the concoction that US gold medalist Thomas Hicks swilled before and after the 1904 Olympic marathon) and some research on hangovers and physical performance, but those weren't really the scenarios I was interested in. I wanted to know whether dropping by the beer tent after an event would harm my recovery. The most relevant research I found suggested that drinking alcohol might impede the body's replenishment of muscle-fuel stores after exercise or perhaps slow the repair of muscle damage. That seemed interesting, but the studies had only examined rugby players and weight lifters. Would the results translate to the masses of us who run or cycle on the weekends?

I started my career as a lab researcher, and I'm still an experimentalist at heart. So I dropped by the Monfort Family Human Performance Research Lab at Colorado Mesa University to propose the idea of conducting our own study to my friend Gig Leadbetter, who was then a researcher at the lab and head coach of the school's cross-country team. Tall, with a lanky runner's build and a goofy smile, Gig has a curious mind and is always open to new ideas. I had a feeling that Gig would be game, since he brewed his own beer and dabbled in winemaking. (He recently retired from CMU to start a cidery.) When I asked him if he'd be interested in doing a study about beer and running, he didn't hesitate. "Let's do it!"

———

A few weeks later, we got together in a conference room with a few of Leadbetter's colleagues and hashed out a study design. Our study's objective was to test whether drinking beer after a hard run would have any effect on recovery, and the first decisions we faced were: how would we define "hard run," and how would we measure recovery? The first question was relatively easy to answer. Because our hypothesis was that the alcohol in beer might alter the replenishment of muscle glycogen, we needed the pre-beer run to be at a pace that would diminish these energy stores. We'd do that by testing the aerobic capacity of our runners ahead of time, then putting them on a treadmill to run for a pace and distance that would deplete their glycogen.

As for the test of recovery, Leadbetter suggested we use a so-called run to exhaustion or RTE. We'd put our volunteers on a treadmill, set the speed to 80 percent of their maximum, and have them run until they couldn't go any longer. This wasn't some sadistic plot that Leadbetter had dreamed up—it's a standard test used in many other exercise studies. Given its widespread use as a recovery measure, it seemed like an easy choice.

With those two decisions made, Gig and his colleagues mapped out the rest of our three-day protocol. If beer (or, more specifically, the alcohol in beer) impairs recovery, we'd expect that runners who drank it after their depleting afternoon run would run out of gas faster in the next morning's run to exhaustion compared to those who didn't drink booze. Beer might also make the exhaustion run *feel* more difficult, so we'd ask runners to rate how hard the effort felt throughout the run. Finally, the alcohol might alter the proportion of fat and carbohydrates burned for fuel, and so we'd also take a metabolic measurement that could capture this too.

Before the study began, we gathered the participants in the lab for a prestudy briefing. Ten of us took part (including me), and our ages ranged from twenty-nine to forty-three. I'd recruited most of the runners from the Mesa Monument Striders Running Club, and we were all moderate drinkers who ran at least 35 miles per week. Leadbetter handed out an instruction sheet and explained how the study would work. He also put us through a bit of pretesting. The protocol called for serving volunteers enough beer to get them to 0.07 percent blood alcohol, just below the legal limit of 0.08 percent for "driving under the influence" in Colorado. The point was to simulate the amount of beer you might drink after a race or hard workout. Leadbetter had used a standard alcohol chart that used body weight to estimate the amount of beer each of us would need to get to the target, but since metabolisms can vary, he invited a local cop to the meeting to make sure we were on track. As the meeting proceeded, we each drank our prescribed amount of New Belgium Fat Tire beer, and then the friendly cop called us up to test each one of us with the Breathalyzer.

It was a good thing Gig called for backup. The chart proved right on the mark for some, but was far less accurate for others. It correctly predicted, for example, that twenty-nine-year-old Daniel needed to drink three-and-a-half beers to reach 0.07 percent. However, it calculated that Bryan, a muscular 149 pounds, would need to drink almost three beers. As he strode to the front of the room and faced the cop for his moment of truth, he didn't seem at all impaired. But when he blew into the Breathalyzer, the number came up to .095 percent. The cop cracked a big grin.

"Oh yeah, you're good to go!" He meant Bryan would be going to jail if he were caught driving like that.

Cynthia, a petite and speedy runner, was only allowed one beer based on the body weight chart, but the Breathalyzer revealed that she really needed almost two to get to the intended level of drunkenness. By night's end, as we met our designated drivers, Leadbetter and his team knew exactly how much to pour.

In the weeks before the experiment, we each underwent a few hours of pretesting to determine our fitness parameters. The study itself took place over three days. The protocol looked like this:

Day 1
Evening: 45-minute run at 75% of max, followed by beer
    and pasta.
Day 2
Morning: Run to exhaustion at 80% of max.
Evening: 45-minute run at 75% of max, followed by beer
    and pasta.
Day 3
Morning: Run to exhaustion at 80% of max.

We began on a Friday evening with a 45-minute run on a treadmill set to a speed that was 75 percent as fast as each of us was running when we'd hit our VO2 max (a measure of how much oxygen you can use when you're going full tilt) during the pretesting. The pace felt swift, but not all-out. After the run, the beer drinking began. Gig and his gang set up a camp stove on the lab's back patio and cooked up a pasta dinner—spaghetti with red sauce, a salad with vegetables from Gig's backyard garden, and garlic bread. As we ate, one of the undergraduate assistants poured each of us runners an individually measured amount of beer. They delivered the beers in clear plastic cups with our names written on them, which sort of made it feel like we were at a classy kegger. Half of the runners drank Fat Tire Amber Ale; the other half received our placebo beer—O'Doul's Amber, a nonalcoholic beer that looks a lot like Fat Tire. The design was double-blind—only the student researcher knew who got which beer, and he secretly poured them in a shrouded room so that neither the runners nor the other researchers knew who was getting which beer.

While we slurped spaghetti and downed our beers on the back patio, we cracked jokes about the placebo beer and took (easy) guesses about which beer we were currently drinking. It didn't feel that different from the kind of refreshments and banter we might have shared after a club run or race.

The morning after the first run, we all returned to the lab for the run to exhaustion—a treadmill run paced at 80 percent of the speed we'd been running when we hit our VO2 maxes in the prestudy testing. It was a pace that felt taxing and made it too hard to converse in full paragraphs, but not over the red line. I couldn't have spoken even if I'd wanted to, because my nose was pinched closed with a clip, forcing me to breathe through a plastic tube hooked up to my mouth so the researchers could measure the gases I was breathing in and out. The apparatus felt foreign and a bit clunky, but it didn't impede my breathing so I put up with it in the name of science. Every three minutes, I rated how hard I was working by pointing to a number on a scale of perceived exertion, and I was supposed to continue running until I couldn't go anymore. During the run, Gig and his team cheered me on as they also measured my heart rate, oxygen use, and respiratory exchange ratios. These measures would give us some insight into whether beer provoked metabolic changes. After this test, I ate a hearty breakfast and talked with the other runners about how weird and difficult the RTE had felt.

Late that afternoon, we returned again for beer run number two. Once again, everyone ran 45 minutes at 75 percent of their maxes, and once again, we ate a big pasta meal and drank whichever kind of beer we didn't get the first time. (This design allowed us to compare each runner's recovery after drinking the alcoholic versus the placebo beer.) The study wrapped up the next morning, when we all returned for the final run to exhaustion. By the end of the weekend, we'd bonded through our shared beers and mutual agony during the RTEs.

With the study done, we anxiously awaited the results. When the analysis was complete, Gig called me up with exciting news. We'd hypothesized that there were three measures that alcohol might alter during the run to exhaustion: ratings of perceived exertion (how hard the run felt), respiratory exchange ratios (a measure of what kind of fuel the body is burning), and the time required to reach the exhaustion point. The results showed no difference between trials for the first two factors, but the third one, time to reach exhaustion, showed a difference worthy of publication.

It turned out that the men in our study reached their breaking point an average of 21 percent *sooner* the morning after they'd downed the alcoholic beer, while our women, on average, kept running 22 percent *longer* during the run to exhaustion following the alcoholic beer. The analysis showed that the difference for the men was not statistically significant, suggesting that the differences in the two trials were within the range that might be expected if beer didn't affect the measures we were testing. If beer had an effect on the men's recovery, it was inconsequential. But the difference between the exhaustion runs for women did reach statistical significance, which meant that we would have been unlikely to see differences like this if beer didn't have an effect. These results gave me license to write a story for *Runner's World*, which had funded our study, that said something like this: "Beer boosts running performance in women!"

If our results were correct, it meant that women could improve their recovery after a hard run by drinking beer. Honestly, what more could a beer-loving runner ask for? It meant that my postrun beers weren't an indulgence, but a *scientifically proven* performance enhancer. For women, at least. The findings also suggested a new surgeon general's warning: "ACCORDING TO SCIENCE MEN SHOULD NOT DRINK ALCOHOL IF THEY NEED TO RUN HARD TOMORROW." Sure, I would have preferred that our study suggested that postrun beers were good for guys too. On the other hand, this result provided an easy argument for making my husband the designated driver. I needed beer for my recovery. He didn't. Our results implied that for women, beer wasn't an indulgence, but part of a smart training plan. I'll drink to that!

———

There was only one problem: I didn't believe it. Trust me—I wanted our study to show that beer was great for runners, really, I did. Yet my experience as a participant in the study left me feeling skeptical of our result, and the episode helped me understand and recognize some pitfalls that I've found to be common among sports performance studies. These problems aren't a matter of deceit or scientific malpractice, at least not usually. Instead, they're the result of difficult obstacles that face any researcher hoping to crack exercise physiology or sports performance with science. The bottom line is that science is hard, and sports science especially so.

My doubts began with the run to exhaustion. We had selected this protocol as our measure of recovery because we thought: Why reinvent the wheel? Putting runners or cyclists on a treadmill or stationary bike and making them go until they can't continue is an accepted methodology used in countless other studies, so we never questioned it. When I asked if drinking a beer after a hard run would impair my recovery, what I really wanted to know was whether it would make me feel cruddy and less able to run well the next day, and the RTE seemed like a reasonable way to measure that.

At least, that's what I thought when planning the study. But my experience as a participant quickly convinced me that the exhaustion test was a lousy way to measure what we were trying to study. For me, the RTE didn't feel like it was testing my recovery from the previous day. Instead, it felt like a test of how much discomfort and annoyance you're willing to tolerate in an exercise lab. You're running at 80 percent of your max: it's hard, but it's a step or two down from all-out. Your legs get a little heavier and your will to continue wanes, but you can't quite reach that totally spent feeling you get after a finish-line sprint. It becomes a mental game—how long can you put up with this feeling of discomfort but not quite total exhaustion? "I kept asking myself—am I truly exhausted, or just sick of this?" Cynthia, the fast and petite runner, told me afterward. We agreed that mostly we were just uncomfortable and bored. I kept feeling an urge to turn up the treadmill so that I could burn the last of my energy in one big spurt; instead, I was forced to let it trickle out, one step at a time. It was a slow, gentle torture.

On the one hand, when I'm tired, my motivation wanes, so

in that sense the RTE probably captures an important psychological aspect of recovery. It's certainly not a worthless test.
But I'm not convinced that it captures the things that matter
in the real world. The problem is that it's done in a lab setting at
a prescribed pace, rather than at a speed we'd naturally select
if given free rein. We couldn't sense how we were feeling and
adjust our pace accordingly throughout the course, as we'd do
on a normal training run or in a race. And the motivation felt
a bit artificial too—there wasn't really anything in it for us, we
were just volunteering for the greater good. The results from
Larry, a tall and lanky marathoner, provide a perfect example
of what I'm getting at. He ran ten minutes and forty-six seconds
longer on the RTE after he drank the nonalcoholic beer than he
did the morning after drinking the Fat Tire. After the results
were tabulated, I asked him if he realized that he'd performed
worse after the real beer. "Yeah," he said. "I probably could have
gone a little longer that time, but I had my daughter with me
and I wanted to get done so we could go home."

Talking to other participants after the study was over, there
was a general sense that the open-ended nature of the RTE gave
the test a sense of arbitrariness, like a contrived game of attri-
tion. It's an exercise unlike any real event I've ever done in real
life, and it's not clear to me how it translates to the things I care
about. Turns out, these concerns aren't unique to us. One assessment of various tests of athletic performance concluded that a
time trial or race of a set distance produced results that had better reliability, validity, and sensitivity compared to tests like
our run to exhaustion that asked people to continue exercising
until they wanted to quit.[2] That's important to know, because if
you're using a test to measure something, you want to know that
any differences you see between one trial and another are not

just part of the normal variation that might happen if you did the test again under identical circumstances.

The important lesson I took away was that it's crucial to ask whether a study is really measuring what it's supposed to and whether that measurement translates to something you care about in real life. In our study, the ratings of perceived exertion—which are essentially just an answer to the question "how do you feel?"—seem more relevant, and on those we found no clear pattern to suggest that beer had an effect.

One strength of our study was that it was randomized and double-blind—runners were randomly assigned to get one or the other beer on the first trial, and neither the participants nor the researchers doing the measurements or analysis knew which RTEs were done after alcoholic beer and which ones were done following the placebo beer. Such a design represents the gold standard for this type of research, but despite all the effort that went into the blinding and the placebos, it was pretty easy for most of us to figure out which beer we'd received. Larry even surmised the brand of the nonalcoholic beer that he'd received. (Turns out, he'd tried them all previously when he'd quit drinking alcohol for a while in hopes of bringing down his marathon time. His times didn't budge, so he went back to the good stuff.)

I've since learned that this blind but not blind issue is a common problem with many studies of recovery tools like icing, sports drinks, and massage, where it's hard to create a convincing placebo. Once you know what you're getting, it's easy for this knowledge to sway your expectations and therefore your performance, even if you're not trying to game the results. If you know you got the alcoholic beer, that may give you license to quit a little sooner on the RTE. On the other hand, if you want to think that alcohol could give you an edge, you might be more motivated to keep going.

During our experiment, I discovered that it's remarkably easy to tip the results in one direction or another. Some of these inadvertent nudges come directly from the researchers themselves. At our prestudy orientation meeting, someone asked Gig how long we should expect the run to exhaustion to take. "Most people last about 20 minutes or so," he told us. After the study, I confirmed that it wasn't just me—this little bit of information primed those of us at the meeting to shoot for running at least 20 minutes. He'd essentially given us permission to stop after that. If he had told me that most people last 40 minutes, I'm pretty sure I would have run at least that long. We weren't supposed to watch the clock, but on my first trial someone forgot to cover the timer on my treadmill, and on the second trial, I kept tabs by looking at a clock on the far side of the lab. It seems telling that only three of our ten participants ran less than 20 minutes on the RTE, and they were the ones who missed the meeting and got private briefings. The RTE was supposed to be open-ended, but this inadvertent nudge had given me a goal to shoot for.

I also had some doubts about our study's timeline. We'd decided to run the experiment over a single weekend, which meant that runners performed four hard runs in less than 48 hours. We had good reasons for doing this—it was easier for participants to commit their time if they were giving up one weekend, rather than two, and the sequential timing also made it easier to get time in the lab and the necessary staffing. Because half of the runners would get alcoholic beer first and half would get it second, we could theoretically control for the accumulated fatigue on the second RTE. But in practice, it's not easy to do this when you have such a small sample size, particularly when the effect you're studying is unlikely to be very large. By the end of the second day, I felt noticeably tired from all the hard running. Was the effect of the beer greater than the fatiguing effects of the study itself?

If alcohol made a huge difference for recovery, then issues like this shouldn't skew things much, but when you have a very small study like ours, it's easy for a few little factors like this to complicate the results, especially if the effect you're looking for is small.

It's easy to explain away any shortcomings, though. When Gig first told me that the women in our study performed better the morning after drinking beer, I was ecstatic. We'd worked hard to design a rigorous study, and I believed in our science. Everything we'd done was carried out with noble intentions. I wanted the study to turn up an interesting result. It's human nature to want your work to succeed, and I'll admit, I not-so-secretly hoped that our study would prove that running and beer were a good mix. In other words, I was primed to believe in (and overstate) our result. My enthusiasm for our study created a credulous spirit that, left unchecked, could have easily overridden any doubt. As the late Nobel Laureate physicist Richard Feynman once said, "The first principle is that you must not fool yourself—and you are the easiest person to fool."[3]

Once you've got your sexy result, it's really easy to come up with a story to explain it. In our case, we came up with some possible reasons having to do with sex hormones and glycogen replenishment rates. These were plausible explanations, but our study didn't measure those factors, so we needed to be careful about pointing to them as the answer. They're a perfectly reasonable hypothesis, but nothing more.

My college anthropology professor taught me a name for narratives people develop to explain their data—"just-so" stories. The name comes from Rudyard Kipling's fanciful animal tales for children, which explain, for example, that the camel got its hump as punishment for being lazy. Just-so stories are appealing because they so perfectly explain the data you've found. That's not because they're true, but because they were *explicitly created to fit the data*. There's nothing wrong with thinking about possible explanations for scientific results—stories are how we put them in context and ascertain how plausible they are. But it's critically important to avoid falling too in love with these untested explanations. A good scientist never loses sight of what's evidence and what's conjecture, but maintaining a clear wall between them can be tricky, because as humans, we're drawn to stories that *feel* true. When the story fits what we want to believe, it's easy to overlook its flaws.

When you take results from a small sample like ours and average them, the just-so story you write can easily obscure the real picture. Our averages told a compelling tale, but when you look at the raw data, it's not as convincing. The individual numbers were all over the place. One female participant ran 74 percent longer after drinking beer, while another went only 16 percent more. On the men's side, one guy ran 32 percent longer after his O'Doul's, while another actually ran a sliver more after the alcoholic beer. Removing the most extreme result from either gender would have altered the answer we derived. Viewed like this, I wondered—are we really seeing a pattern here, or just forcing a line through our messy data?

One important limitation of our study was its size. Small studies are generally less reliable than larger ones, because they're less likely to constitute a representative sample, and they're known to have a bias toward showing a positive effect for the thing that they're testing.[4] A significant result from a small study is more likely to be a false positive than a significant result from a large study. In a paper published in 2012, psychologists calculated the likelihood of obtaining statistically significant results and showed that it's easier to meet this goal by doing five small studies with twenty participants each rather than one study with one hundred people.[5] Even though the false positive rate for each individual study is only 5 percent (assuming they use the standard threshold for significance), a series of five small studies gives more opportunities for bias than a single larger study, and those 5 percent false positive rates add up to a nearly 23 percent false positive rate when you put the five studies together. A positive result from a small study is an interesting start, but to trust it, it needs verification, preferably in a larger sample. Our study was intriguing, but it was too small for us to be confident that beer's benefits for women were real or that it really had no important effect for men.

———

It may seem impolite to look for flaws or sources of uncertainty in a study like ours, which was done in good faith with the best intentions. But the most important questions scientists should ask about any study are: How could this result be wrong? What are the things we're sure about, and which things remain less certain? The purpose of these questions isn't to tear down the work but to learn as much as possible from it. A single study can never give the definitive last word, because science must always remain open to new evidence. Our study was just one small piece of the beer puzzle, and thinking critically about it offers an opportunity to figure out what the next study should do and how it can improve and build on this one. To really trust the results, we needed to repeat the study and improve the methods to verify that what we'd seen in the first study held up.

Since our study, there's been a little bit more research, but there's still no definitive answer on how (or whether) a beer or two after a workout will influence recovery. Like many research questions, the most accurate answer probably includes the phrase "it depends."

A 2014 study led by researchers in Australia tested the effects of alcohol consumed following a hard bout of weight training and cardio. The results showed that when the subjects drank alcohol after the strength training, a kind of muscle repair that takes place after muscle-damaging exercise like my run down Mt. Garfield was less than when they'd just ingested protein after the exercise.[6] This was true even when they imbibed the alcohol along with protein or with carbohydrates. It's an interesting result, but the study was small—only eight subjects, all men—and they used a dose of alcohol (1.5 grams per kg of body weight) equivalent to about seven beers for a 150-pound person. That's more like a heavy night of drinking, not a happy hour refreshment.

Matthew Barnes, an exercise researcher at Massey University in New Zealand, has done numerous studies on alcohol and exercise recovery.[7] Following strenuous strength training, the exercised muscle's strength is typically diminished for a period of up to 60 hours while the muscle rebuilds and repairs itself. Barnes's work found that this effect in muscles was heightened when exercisers drank 1 gram of alcohol per kilogram of body weight (the equivalent of about 5 drinks for someone weighing 150 pounds) afterward, but unaffected if the amount of alcohol consumed was half of that, or 2.5 drinks for a 150-pound person. "One or two beers may be okay, but drinking to excess isn't wise. A lot of people will say that's common sense, but now we have the research to back it up," he says.

Although the research suggests that having a little alcohol probably won't noticeably harm recovery, how much is too much isn't clear, Barnes says. "We still don't really know the relationship between dose and effect, and we don't know anything about the timing." What we know about alcohol and recovery at the moment comes from small studies, most with ten or fewer participants, so it's hard to draw definitive conclusions. As for the gender difference we (maybe) saw, Barnes says that some studies have suggested that estrogen can offer a protective effect against exercise-induced muscle damage, and since alcohol may increase the production of estrogen, this may explain why women in our study may have gotten a boost from beer. It's a pretty good just-so story, but while we wait for more studies, that's all it is.

Given what we know right now, Barnes says, the best recommendation is to make sure you quench your thirst with some water or nonalcoholic beverage after exercise and have a decent meal with some carbohydrates and protein along with or before the beer. "As long as that's been done, then a small amount of alcohol is probably not detrimental," he says. Cheers to that!

———

In the end, my investigation into beer left me where I had started, at least as far as what I thought about postexercise beer (go ahead, in moderation and with some common sense). At the same time, it completely upended my thinking about the scientific process and what studies can tell us about the effectiveness of various approaches to improving or expediting exercise recovery. What I learned in the beer study guided me as I started to investigate the strange world of recovery methods.

For one thing, I discovered that it's not enough to ask "Does this thing work?" First, you have to start with more fundamental questions: How would we know if it's working? What are the benefits this gizmo or ritual is supposed to deliver, and how would we measure them? If the proof is coming from something measured in a lab, do those numbers translate into meaningful differences in real life? As I found out during the run to exhaustion test, just because you can measure something doesn't mean it's answering your question.

Another important lesson from the beer experiment is how naive I'd been to even hope that a single study could deliver *the* answer. Those eureka moments where a researcher in a lab coat discovers some incredible new thing that changes *everything* (and shouts, "By god, it works!" while shaking a fist) are the stuff of sci-fi movies. But in real life, they're few and far between. The boring truth is that most science is incremental. Neither our experiment nor the other beer studies in the scientific litera-ture provided the definitive last word about whether beer could hurt or harm recovery, but taken together they offered a picture that represents the best answer that we have—at least until new studies come along to add even more nuance and detail. As I delved further into the research on recovery methods, I had to make peace with some uncertainty. Sometimes the best answer to the question "Does it work?" was: maybe.

I also learned to keep my eyes open to ways that researchers (and athletes) might unwittingly fool themselves into think-ing that they'd found some recovery magic, especially if it was something they really wanted to believe. I knew this could hap-pen, because I'd nearly done it myself. (To be honest, I still cling to a small hope that beer is performance-enhancing.) Sprinkle an appealing idea with a dash of science, and it can seem more powerful or true than the evidence really shows. But good luck overturning an idea once it's become part of sporting lore. That was a lesson I would learn soon enough.

2

Be Like Mike

In the early 1990s, Gatorade ran a television commercial featuring Michael Jordan that would become an icon in the advertising world and inspire striving athletes across North America. The ad was called "Be Like Mike," and it featured slam dunks by Jordan interspersed with footage of kids shooting hoops and, of course, Jordan and other happy people drinking Gatorade from the iconic glass bottles with the fat bottoms and tapered necks that the drink was sold in at the time.[1]

Stuart Phillips remembers that ad campaign well. As an aspiring athlete, he, too, wanted to be like Mike. "Michael Jordan drank Gatorade, so I drank Gatorade," says Phillips. Despite guzzling the sports drink, Phillips never did make it to the pros, but instead grew up to become the director of the Centre for Nutrition, Exercise, and Health Research at McMaster University in Hamilton, Ontario. The Jordan ad taught him a lesson about the power of marketing, though: "If you can get an endorsement from an athlete that everybody recognizes, then who needs science?"

Scientific facts don't sell products; stories do. If you're aiming for blockbuster sales, nothing beats a compelling narrative, and no one has mastered the art of storytelling better than Gatorade. Jordan was already a basketball superstar by the time Gatorade came calling, and the public was eager to experience something of his greatness. Enter Gatorade—Michael Jordan drank it and young Stuart Phillips could too. To drink Gatorade wasn't just to mimic a sports hero, it was to imagine a causal relationship—Jordan drank Gatorade, then made all those slam dunks, so the one must have had something to do with the other. Psychologists call such thinking the "illusion of causality," and it's so powerful that it has spawned an entire genre of advertising—the celebrity endorsement. No one would care that a pro athlete uses a particular product if it didn't somehow appear that the item played some role in that star's success. The Irish have a saying, "An umbrella accompanies the rain but rarely causes it." The same could be said of product endorsements and athletic greatness. Still, our minds are quick to connect the dots in the wrong direction.

———

The age of the athlete-endorsed sports drink began on a Florida football field in the mid-1960s. Back then, most coaches and athletes didn't give much thought to fluid replacement during practice or competition. In some instances, athletes were even counseled to avoid drinking close to a workout, lest they upset their stomach. But in 1965 a University of Florida football coach came to Dr. Robert Cade and his team of university doctors,[2] complaining that his players were "wilting" in the heat. (He also wondered why his players never urinated during games.) After some investigation, Cade and his colleagues concluded that two factors were causing the players to fall victim to the

heat—they weren't replenishing the fluids and salts they were
sweating out, nor were they restoring the carbohydrates their
bodies were burning for fuel.

Cade figured he could solve the problem by helping players replace these lost resources, so he stirred together some
sodium, sugar, and monopotassium phosphate with water to
create a drink soon dubbed Gatorade, after the University of
Florida's nickname: the Gators. Players complained that the
concoction tasted "putrid" so, at his wife's suggestion, Cade
added a bit of lemon flavoring to make the beverage more palatable. Legend has it, the drink turned the struggling Gators
football team around. They finished the season with a winning
record, and in 1967 the team won the Orange Bowl for the first
time in school history. Ray Graves, the University of Florida
football coach from 1960 to 1969, later recounted his team's
win against Georgia Tech in the 1967 Orange Bowl: "After the
game, Georgia Tech coach Bobby Dodge said to me, 'We didn't
have Gatorade, that made the difference.' I believed that then,
and I still believe that to this day." Other teams took notice of
the newfangled beverage, and in 1967 Cade and the University
of Florida signed an agreement with canned goods company
Stokely-Van Camp to produce Gatorade commercially.[3] Orders
for the drink poured in.

What followed was a national campaign to sell the public
on the idea that exercise caused dehydration, the cure was
Gatorade's specially developed drink, and this tonic was critical for sports performance—it was created by a doctor and
tested in studies, after all. "Gatorade is different than juices
or fruit drinks or soda pop or water. Research men created
it; thousands of athletes swear by it," announced the narra-
tor of a 1970 television ad that ended with the product's then-

slogan: "Gatorade Thirst Quencher—the *professional* thirst quencher."

"It's one thing to have a great marketing plan. It's another to execute it to near perfection, as those who worked on Gatorade's brand have done," wrote Darren Rovell in his 2005 book, *First in Thirst*. One of the brand's early print advertisements boasted that Gatorade was absorbed twelve times faster than water (a claim walked back in 1970,[4] after Ohio State team doctor Robert J. Murphy challenged it at a meeting of the American Medical Association).

In a stroke of genius, Gatorade turned the drink's sodium, phosphorus, and potassium into a special selling point by rebranding these ordinary salts with their scientific name—"electrolytes," which is simply the scientific term for molecules that produce ions when dissolved in water. Your body maintains some reserves of these vital ions that it can tap into as needed to keep your body's fluid and salt balance in check. We do lose electrolytes through sweat, but even when you exercise continuously for many hours you will simply correct any losses via your normal appetite and hunger mechanisms. (You've already experienced this if you've ever had a hankering for a salty snack.)

There's no reason to salt your water (or your beer, as one Australian researcher did in an ill-fated attempt to make a more hydrating brew).[5] You can replenish your electrolytes with food instead. In a study involving ten men who were trained cyclists or triathletes, researchers found that it didn't really matter whether they drank plain water, a sports drink, or a milk-based beverage after an hour of hard exercise.[6] As long as they drank some liquids along with a meal, they restored their fluid levels just fine. The research suggests that knowingly or not, people naturally self-select foods that compensate for whatever salts or minerals they lost from sweat. Even if you need to replace salt, that doesn't mean you have to drink it.

Gatorade may not have been the first to use this term, but they're the ones that landed electrolytes in the public lexicon. Because electrolytes weren't (yet) a household word in the early days of sports drinks, people could mistake them for special compounds that needed to be taken in a sports drink's magical formula if you were to replenish the fluids lost during exercise. Other products boasted about electrolytes, but Gatorade quickly became the most popular.

Sales boomed, and in 1983, Gatorade was acquired by the Quaker Oats Company. The NFL signed a deal that same year to make Gatorade the league's official sports drink, and in 1985 the Gatorade Sports Science Institute was founded to promote the study of hydration and nutrition for athletes, research that also happened to make for great marketing. Conveniently, the studies that came from the GSSI could be used to support the product's claims. "We test Gatorade in laboratories. We test it at major universities, with sports science experts, on sophisticated scientific equipment with names that are longer than this sentence. What does it prove? Gatorade works," read a 1990 magazine ad.[7]

Early advertisements presented thirst as the problem that Gatorade was designed to solve, but as the GSSI's research program progressed, the emphasis moved to a more clinical concept of hydration and the notion that thirst was not a good indicator of whether an exerciser was drinking enough. "Unfortunately, there is no clear physiological signal that dehydration is occurring, and most athletes are oblivious to the subtle effects of dehydration (thirst, growing fatigue, irritability, inability to mentally focus, hyperthermia)," wrote GSSI cofounder Bob

Murray in one report.[8] Instead, athletes were advised to drink according to scientific formulas. A 2001 Gatorade ad depicted the glistening torso of a runner with the race number 40 pinned to her shorts and the words, "Research shows your body needs at least 40 oz. of fluid every hour or your performance could suffer."[9] That's the equivalent of five 8-ounce glasses of liquid, which means a runner finishing a marathon in a fast three hours would need to drink 15 glasses of fluid along the way. Gulp.

———

Gatorade wasn't alone in promoting the benefits of drinking before, during, and after exercise. Other sports drink manufacturers, such as the drug company GlaxoSmithKline (Lucozade Sport), also pointed to science when marketing their products. Lucozade, for example, established a "sports science academy" to promote its drink. Together, these campaigns fostered the idea that exercise depletes your fluids and electrolytes (which, remember, is just a fancy name for salts) and that special measures are required to make things right again.

It was no longer sufficient to simply drink some water and eat a meal after exercising. The idea these marketing campaigns fostered was that physical activity created extraordinary nutritional needs, and these specially formulated beverages were the best way to meet them. This was science speaking.

Sports doctors were also urging athletes to drink. The American College of Sports Medicine (ACSM), a professional organization of sports science experts (which happens to receive financial support from Gatorade[10]), put out a consensus statement in 1996 recommending that "during exercise, athletes should start drinking early and at regular intervals in an attempt to consume fluids at a rate sufficient to replace all the water lost through sweating (i.e., body weight loss), or consume the maximal amount that can be tolerated."[11] The message coming from experts was that athletes needed to replace the fluids they lost during exercise, lest their performance and health suffer.

In the wake of all this promotion, sports drinks have become a multimillion-dollar business. Old standbys like Gatorade and Powerade (Coca-Cola's contribution to the market) now compete with products with names like Propel, Accelerade, Maxade, and Cytomax. Sports drinks have become as ubiquitous as soda and bottled water. But when a team of medical researchers trained in the evaluation of scientific findings had a look at the research underpinning the boom in sports drinks, they reached a startling conclusion. "As it turns out, if you apply evidence-based methods, 40 years of sports drinks research does not seemingly add up to much," Carl Heneghan and his colleagues at the University of Oxford's Centre for Evidence-Based Medicine wrote in a 2012 analysis published in the *British Medical Journal* (*BMJ*).[12] When Heneghan's team gathered and examined all of the available evidence on sports drinks (they even consulted sports drink manufacturers to ask them for their supporting studies, though not all complied), they found what amounted to a bunch of preliminary or inconclusive evidence packaged as more definitive proof.

The first, almost universal, problem among these studies was that they were too small to produce meaningful results. "Small studies are known to be systematically biased toward the effectiveness of the interventions they are testing," Heneghan and his colleagues wrote.[13] Out of the 106 studies they analyzed, only one had more than 100 subjects, and the second largest study used only 53 people. The median sample size? Nine. Most of these studies were about as definitive as my beer experiment.

Another common shortcoming was that the studies were often designed in a way that almost assured that they'd find a benefit from sports drinks. Some of the study setups verged on "comical," says Deborah Cohen, an investigations editor at *BMJ* who was involved in the project and wrote a summary of the findings.[14] She recalls one study where volunteers fasted overnight and then one group was given a sports drink, which contains water, salts, and sugar, and the other received water. "People who were given the sports drink fared better. Well, no shit," she told me. If you haven't had any food in 12 hours and then you get a bit of sugar, of course you'll perform better than the people still running on empty. But to say that this means that the sports drink is superior to whatever a normal person would consume leading up to or during exercise just isn't generalizable, she says. "Who starves themselves overnight and then goes to perform some exercise?" And yet the *BMJ* investigation found that this type of study design is surprisingly common among tests of nutritional products. Instead of comparing a product to what athletes would otherwise consume, they compare some new nutritional product against exercising on empty. Cohen argues that's not a fair test of the product's benefits in real-world conditions.

Some of the dazzling powers that sports drinks display in the studies touted by their makers may be nothing more than the placebo effect. When people volunteer for a study to test a new sports drink, they come to it with an expectation that the product will have some performance benefit. Studies use a placebo group to factor out such effects, but a placebo only controls for these expectations when it's indistinguishable from the real deal. So it's telling, Cohen says, that studies using plain water for the control group found positive effects, while the ones that used taste-matched placebos didn't.

The *BMJ* analysis also concluded that many of the measures made in these studies may have looked good on paper, but aren't things that matter when it comes to real-world performance. Very few athletes ever compete in events where the purpose is to keep going until you can't. Instead, most of us care more about the answers to questions like, Does it help me perform better or feel less fatigue? "Worryingly, most performance tests used to assess sports drinks have never been validated," Heneghan reports, and some of them, such as tests to exhaustion, are known to produce highly variable results, which means that repeating the test under the same conditions can turn up different numbers. I noted this problem myself in our beer study, where the run to exhaustion felt more like a test of concentration and desire than of recovery. Sure, the studies are churning out numbers, but those numbers may not answer the question at hand. As I learned in our experiment with beer, sometimes the numbers aren't measuring the things you care about.

Heneghan and his team concluded that claims about sports drinks rely on small studies with comparison groups that rig the studies in favor of the products being studied, a lack of rigorous blinding so that participants were likely nudged to perform better while taking in the sports drinks, and measurements of effectiveness that might not be meaningful in real life. Add to that statistical sleights of hand that inflate the benefits of the drinks (for instance, one study increased the benefit of carbohydrate drinks from 3 percent to 33 percent by excluding a segment of the test from the analysis), and sports drinks don't come out looking so impressive.

When the report came out, some sports science experts blasted Heneghan and Cohen's report as unnecessarily rigid, because they set their standards based on the conventions of clinical medicine rather than sports science, where, for instance, small sample sizes are common. What standards should be used for assessing evidence is an important debate, but there's another crucial issue here. The marketing around sports drinks rests on a fundamental, seemingly scientific premise—that even minor dehydration raises health risks and hinders athletic performance and recovery—but this idea appears overstated. It's more marketing than science.

———

Amby Burfoot was a participant in one of the early studies commissioned by Gatorade,which exercise physiologist David Costill conducted at Ball State University in March 1968.[15] Costill is widely considered one of the pioneers of sports science research, and Burfoot arrived at his lab in Muncie, Indiana, having just run at an NCAA indoor track meet in Detroit, where he'd been beaten, badly, by track legends Gerry Lindgren and Jim Ryun. Burfoot had never run on a treadmill before (the contraptions were virtually unheard of at the time), but over the course of the next several days, he'd become well acquainted with one as Costill put him through the grueling protocol: three separate two-hour treadmill runs, completed at 70 percent of his aerobic capacity, which for Burfoot equated to a 6-minute-per-mile pace.[16] During one of the runs, Burfoot drank nothing, on another he drank water, and on the third occasion he was given Gatorade at regular intervals. "It was for science, and Dave was fun and solicitous," Burfoot says.

Burfoot was an elite distance runner who'd worked his way up

through the ranks in New England during the 1960s. "Because
the Boston Marathon started at noon, every other race director
thought they should start their race at noon. It didn't matter if
it were the Fourth of July or Labor Day—we started at noon,"
Burfoot told me. Even for events that took place in the summer
humidity and temperatures of more than 90°F, "Nobody had
water stops, nobody had bottles or belts or anything like that,"
he says. "The old road-running veterans said that drinking—
especially cold water—would give you stomach cramps, and
therefore we did not drink at all."

During Costill's study, "I felt distinctly best when I drank
nothing, because that was what I was accustomed to, and it
meant I didn't have any sloshing in my stomach," Burfoot says.
During the runs with water or Gatorade, Burfoot was handed
vials of fluid every ten minutes. "It was just horrid. I remember
my stomach feeling like the Pacific Ocean," Burfoot says. "Every
time they'd hold out the beaker I'd groan." During these runs,
Costill measured the athletes' body temperature as well as how
much of the fluid their bodies were absorbing. This last measure
was taken by threading a plastic tube down their nostrils and
into their stomachs. "He told me, pretend you're swallowing
spaghetti," Burfoot says. The study showed that Burfoot's body
temperature rose more when he completed the run without
drinking, even though he felt better under that condition than
when he'd drunk fluids. But if the rise in his body temperature
hurt him, he couldn't feel it, Burfoot says.

Shortly after the study, Burfoot and some of the other
study participants ran the Boston Marathon, and Costill was
there to measure their body weight before and after the event
to track how much fluid they'd lose. (He'd also made predic-
tions about how the study participants would fare in the race,

and sealed them in envelopes to be opened afterward. Costill had predicted that Burfoot would finish last or second to last among the study's runners.) "There was water that we'd grab and pour over ourselves, but we didn't drink it," Burfoot says. People along the course would hand out slices of citrus fruit. "The orange slices would taste so good and sweet, but every once in a while you'd get a lemon and you'd just pucker up," he recalls. Burfoot won the Boston Marathon that day without drinking a single sip, and he lost nearly ten pounds, according to Costill's bathroom scale.

By the current ACSM guidelines, which warn against losing more than 2 percent of one's body weight through sweat during an event, Burfoot triumphed at the Boston Marathon while foolishly dehydrated.[17] In that sense his win was unremarkable. Ethiopian Haile Gebrselassie lost nearly 10 percent of his body weight during his win at the 2009 Dubai marathon, and he was the marathon world record-holder at the time.[18] Although hydration guidelines instruct athletes to drink according to how much weight they're losing through sweat and respiration, "drinking according to the dictate of thirst throughout a marathon seems to confer no major disadvantage over drinking to replace all fluid losses, and there is no evidence that full fluid replacement is superior to drinking to thirst," the study's authors wrote. Furthermore, athletes who lose the most body mass during marathons, ultramarathons, and Ironman triathlons are usually the most successful, which suggests that fluid losses are not as tightly linked to performance as sports drink makers claim. Instead, the results imply that there must be some tolerable range for dehydration that doesn't impair performance. If anything, the results suggest that some moderate amount of fluid loss might enhance performance, perhaps by lightening the load that an athlete needs to carry over the given distance.

If performance is what you care about, then maybe body temperature and fluid loss are the wrong measures. Subsequent studies have shown that as we get hot and sweat, body temperature rises a little, but then it reaches a new, slightly higher equilibrium. What was the more important finding in Costill's study of Burfoot on the treadmill—that Burfoot's body temperature remained lower when he was forced to drink more water or that he felt a lot better without all that sloshing in his stomach? We don't judge marathoners' performance by the amount of fluids they lose during the race, but by where they finish in the standings. Lab tests can advance scientific knowledge, but they can also direct our attention to the things easily measured, rather than the things that really count.

———

Exercise scientist and physician Tim Noakes was a believer in the dangers of dehydration until two separate experiences left him questioning what he thought he knew.[19] First, Noakes was involved in a study examining participants in a four-day canoe race. During a particularly rough day, one of the paddlers lost all of his drinking water when it washed overboard as he went through some breakers. Despite having canoed about 50 km without drinking, the paddler's body temperature hadn't become elevated, as the dehydration theory would have predicted. "We weighed him, and he'd lost about eight or nine pounds, but his body temperature was normal and I thought, oh my gosh—body weight loss has nothing to do with body temperature," Noakes says. This was a lightbulb moment, because conventional wisdom held that one of the reasons that dehydration was (supposedly) so dangerous was that it put people at risk for heatstroke, and this finding contradicted that assumption.

The canoe study prompted Noakes to reconsider the idea that maintaining full hydration was essential to staving off heatstroke. Then in 1981, a runner wrote to Noakes describing a strange experience she'd had at that year's Comrades Marathon—a famous 90-km ultramarathon in South Africa. It was the first time that the event had provided drink stations every mile of the 56-mile course, he says, and this runner wrote to say that she'd begun feeling really strange about three-quarters of the way through the race. Her husband pulled her off the course and delivered her to the medics. The first responders assumed she was dehydrated and gave her two liters of intravenous fluid, after which she lost consciousness. She had a seizure on the way to the emergency room. At the hospital, doctors discovered that her blood sodium concentration was dangerously low. The ultimate diagnosis was a medical condition called "water intoxication" or hyponatremia—too little sodium in the blood. Contrary to what the medical crew at the race had assumed, the runner wasn't dehydrated—she was *overhydrated*.[20] She'd drunk so much fluid that her blood sodium had become dangerously diluted to the point of hyponatremia. Low blood sodium causes cells in the body to swell, and when this happens in the brain the results can be deadly.

Her case was the start of a trend. In 1987, sixteen participants in the Comrades Marathon ended up in the hospital with hyponatremia, and in 1988, Noakes and his colleagues stood by at Comrades, ready to study any athletes who developed the condition. Eight runners collapsed with hyponatremia that day, and Noakes's group found that they had overloaded on fluid by drinking between 0.8 to 1.3 liters per hour. Noakes's calculations showed that it was very easy to become overhydrated by following the guidelines of the time.[21]

Noakes has built a reputation as a loud contrarian on issues ranging from the science of fatigue to the healthfulness of high-fat diets, so it's not surprising that he was one of the first and also loudest voices on the issue of overhydration (the guy wrote a whole *book* about it).[22]

Yet Noakes is far from alone in worrying that the rush to prevent dehydration may have put exercisers at risk of the far more serious condition of water intoxication. In 1986, another research group published a paper in the *Journal of the American Medical Association* describing the personal experience of a medical student and a physician who'd become stuporous and disoriented during an ultramarathon they'd both run.[23] The men were diagnosed with hyponatremia, and they concluded that they'd developed the condition by drinking too much.

There's never been a case of a runner dying of dehydration on a marathon course, but since 1993, at least five marathoners have died from hyponatremia they developed during a race.[24] In addition, scores of other athletes have become gravely ill but survived. In 1998, Kelly Barret, a forty-three-year-old pediatric dentist and mother of three, became the first runner to die in the Chicago Marathon, and it was likely hyponatremia that killed her.[25] The condition has killed military recruits and several young football players too. At the 2002 Boston Marathon, Cynthia Lucero collapsed before the finish. The twenty-eight-year-old, who'd been running the race to raise money for cancer patients, had been carefully drinking lots of fluids along the course, and she developed a case of hyponatremia so severe that it killed her. The same day at the Boston Marathon, researchers from Harvard Medical School took blood samples from

488 marathoners after the finish. The samples showed that 13 percent of these runners had diagnosable hyponatremia, and three of the runners had critical cases of the condition.[26] German researchers similarly took blood samples from more than a thousand finishers of the Ironman European Championship over multiple years and found that 10.6 percent of them had hyponatremia.[27] Most of the instances were mild, but nearly 2 percent of the finishers had severe or critical cases. Although the findings indicate that it is still a rare condition, what makes them especially concerning is that the early symptoms of hyponatremia are very easily confused with those of dehydration—weakness, headache, nausea, dizziness, and lightheadedness.[28]

How did hyponatremia become an affliction of athletes? In retrospect, it may come down to an error of shifted priorities. In the wake of Gatorade's massive success, sports drink makers turned to science to promote their products, and researchers focused their measurements on the things that were easy to measure—body temperature and sweat losses. Based on an idea that dehydration must be a risk factor for heatstroke, attention moved to replenishing fluid loss. The new sell became that hydration was key to preventing deadly heat illnesses like heatstroke.

The problem is that while dehydration can indeed be associated with heat-related illnesses, it's not a universal factor. "You don't have to be dehydrated to get heatstroke," says Samuel Cheuvront, a research physiologist at the US Army Research Institute of Environmental Medicine. Exertional heatstroke can happen pretty fast when someone exercises hard in the heat (or even in the cold)—too quickly for dehydration to happen.[29] Sure, dehydration may increase the likelihood of heatstroke, but it's not usually a cause, Cheuvront says. When he and his colleagues analyzed 20 years of heatstroke data from the military, they found that only 20 percent of cases had dehydration associated with them. "The majority of the time, dehydration is not causing the heatstroke and may not even be related," he says.

But fluid losses are relatively easy to gauge by hopping on the scale before and after exercise, and they are easier to do something about than the hot temperatures that an athlete might be facing. Dehydration also seems to affect performance measures, at least the ones that researchers were gauging in the lab, and that made the case for its importance seem even more compelling. So you've got well-meaning researchers looking for ways to help athletes cope with the heat, and they find a related, but different, problem that over time becomes the new center of attention. Good intentions, nudged, perhaps, by commercial interests, come together to create a paradigm that holds hydration as paramount. Athletes are told to drink before they're thirsty, because they lose fluids through sweat before their thirst will compel them to drink, and even small amounts of fluid loss seem detrimental to health and performance.

The problem with this model of hydration is that it overlooks basic physiology. It turns out, your body is highly adapted to cope with losing multiple liters of fluid, especially during exercise. When you exercise, you lose fluid and salts through sweat, and that translates into a small change in what's called your "plasma osmolality"—the concentration of salts and other soluble compounds in your blood. You need enough fluid and electrolytes in your blood for your cells to function properly, and this balance is tightly regulated by a feedback loop, says Kelly Anne Hyndman, a professor of medicine at the University of Alabama at Birmingham and leading expert on kidney physiology. When you sweat, your brain senses the cor-

responding rise in plasma osmolality and directs the release
of antidiuretic hormone (ADH), which prods the kidneys to
activate aquaporins, which are like tiny straws that poke into
the kidneys to draw water back into the blood. "It's a pathway to
conserve water," Hyndman says. As your body reabsorbs water,
your plasma osmolality returns to normal, your brain senses
the change, and it shuts down ADH. This feedback loop is finely
tuned to keep plasma osmolality in a safe range. Even a tiny
drop in electrolytes will activate this system to keep your fluid
balance in check. "People always worry they're going to be dehydrated, when the reality is, it's much easier to over-hydrate,
because our bodies are so good at conserving water," Hyndman
says. "Being a little dehydrated is not a bad thing. Our bodies
can handle it."

Athletes who develop hyponatremia during exercise usually get there by drinking too much, because they've been conditioned to think that they need to drink beyond thirst, says
Tamara Hew-Butler, a professor of exercise science at Oak-
land University who is also the lead author of several papers
and a consensus statement on hyponatremia.[30] Even if you
don't drink anything (which she does *not* recommend), your
blood sodium levels will rise in response to sweat losses, and
as a result, your body will shift fluid into the blood to maintain your fluid balance, Hew-Butler says. The same feedback
loop that calls in the aquaporins also activates your thirst,
because the aquaporins kick in before you feel thirsty. "You
don't have to drink above thirst—you'll be fine!" she says. Just
as sleepiness is your body's way of telling you it's time to sleep,
thirst is how your body ensures that you seek fluids when you
need them. No one tells you to sleep before you're tired, and
unless you're in a situation where you can't drink for a pro-

longed period, there's no sense in drinking before you feel thirsty either. Your body is a finely tuned machine that has evolved to adapt to changing conditions, and it's not usually necessary to try and outsmart it.

You can also forget those pee charts that look like paint swatches for urine, and ignore anyone who says that yellow pee is a sign you need to drink more water. If you think about hydration from the standpoint of what's going on inside your body, it's easy to see why urine hue isn't helpful. The color of your pee is essentially just a measure of how concentrated your urine is. If it's got more waste than water, it looks dark, and if it's mostly water, it's light or almost clear. But that's not what's important. What you really want to know is what's going on in your blood, and your urine can't tell you that.[31] Dark pee might mean that you're running low on fluid, but it could also mean that your kidneys are keeping your plasma osmolality in check by conserving water. Very light or clear urine just means that you've drunk more water than your body needs, and that's not necessarily a good thing, especially right before an event.

It turns out that using body weight to determine your hydration status can also be misleading, says Cheuvront. Someone running a marathon burns something like 2,000 or 3,000 calories. They're losing weight in the process, but it's not just water. Even as much as a 3 percent loss of body weight may not translate into any meaningful water loss, in part because burning fuel such as fats and carbohydrates releases water as a breakdown product.[32] Someone who has lost 4 percent of their body weight by the finish line might only be something like 2.5 percent dehydrated, and probably didn't hit that last point until the final miles, Cheuvront says. In events lasting longer than two hours, weight losses could overestimate fluid losses by as much as 10 percent.

Because of the way the body adapts to fluid loss, the common advice to drink lots of fluids in advance of a big event like a marathon may actually backfire. If you drink a bunch of excess water leading up to a competition, you prime your body to become less adept at holding on to precious fluids, says Mark Knepper, chief of the Epithelial Systems Biology Laboratory at the National Heart, Lung and Blood Institute. When you're very hydrated, your body doesn't need to activate many aquaporins, and over time it reduces the number in reserve, meaning that you'll have fewer of these water straws at the ready when you need them. On the other hand, if you have practiced conserving water by waiting until you're feeling pretty thirsty to drink, your body will adapt by directing more aquaporins to stand by. Instead of prehydrating, Knepper says it might actually be better for athletes to practice conserving water in training and continue to simply drink to thirst before an event, rather than trying to top off their fluids.

For athletes doing noncontinuous bouts of exercise like CrossFit or soccer, drinking to thirst in between sets or at halftime is an easy way to keep fluids in balance. But it turns out that even extreme endurance athletes don't necessarily need to drink early and often. The Ultra Sports Science Foundation has published a set of guiding principles for hydration, directed at athletes who compete in long-distance events like ultramarathons or Ironman triathlons.[33] Written by Martin Hoffman, an ultramarathoner and researcher at the University of California, Davis, the guidelines instruct athletes to drink to thirst and to expect some weight loss during exercise. "We did a meta-analysis of studies that have looked at drinking to thirst versus programmed drinking and the findings are that drinking to thirst does not impair performance relative to drinking more than that," Hoffman told me. Losing some weight during a pro-longed event like a 100-mile run or other ultra event is to be expected, he says. If you finish an ultra endurance event at the same weight that you started, you've overhydrated.

Hoffman's guidelines also recognize that dehydration is rarely a cause of heat illness and say that most muscle cramps are not caused by electrolyte loss. (The latest science on cramps suggests that they have more to do with neuromuscular fatigue than with hydration or electrolytes, Hoffman says.) Despite the proliferation of electrolyte supplements and salt tablets aimed at endurance athletes, sodium supplementation is not neces-sary during prolonged exercise "even under hot conditions for up to 30 hours," according to Hoffman. The sodium that an ultramarathon racer consumes in the typical race diet pro-vides enough salt to avoid any fluid-related problems due to salt depletion, he says, and excessive salt consumption during exer-cise doesn't help and may even make overhydration and hypo-natremia more likely. Electrolyte supplements "have become another easy way for people to make money off of vulnerable athletes— by selling them something that costs virtually noth-ing to make," Hoffman says.

———

If our bodies are so good at adapting to moderate fluid loss and letting us know when we need to drink, why are there still so many messages out there urging us to drink before we feel thirsty? The current ACSM, the National Strength and Condi-tioning Association, and National Athletic Trainers Associa-tion guidelines have been updated to warn about hyponatremia, but they still promote the idea that thirst is a poor indicator of hydration and that more than a 2 percent body weight loss should be avoided. An obvious explanation for this is that most of what we hear about hydration comes from companies and researchers with a vested interest in making it all seem complex and highly scientific. (The ACSM, NSCA, and NATAA all receive funding from sports drink makers, and so do some of their members.) If it were as simple as just drinking to thirst, you wouldn't need expert advice or scientifically formulated products like Gatorade or the TB12-branded electrolytes promoted by New England Patriots quarterback Tom Brady. Some companies are now marketing individualized hydration monitors that promise to tell exactly how much fluid you're sweating. But these gizmos don't answer the important question, which is what's happening in your blood and whether drinking more will enhance your performance or recovery.

Yet everywhere I look, it seems that people are telling me to drink more water. In his best-selling 2017 book, *The TB12 Method*, Brady presents his magic hydration formula—drink at least one-half of your body weight in ounces of water every day. "At 225 pounds, that means I should be drinking 112 ounces a day, minimum," he writes. (Brady also contends that "the more hydrated I am, the less likely I am to get sunburned," a claim disputed by scientists.)[34] His drinking formula isn't outrageous, but it's not really necessary either.

From a biological perspective, it's hard to imagine that the human body is so delicate that it can't function properly without scientists (or football stars) swooping in with calculators to tell us how to keep it running properly. "You have to trust your body," says Knepper, the National Heart, Lung and Blood Institute expert. Humans have evolved to survive exercising without chugging water or sports drink on some rigid schedule. "You get clues about what you need if you listen to your own body. You don't have to know chemistry to survive." I've often noticed that water tastes especially delicious when I'm thirsty, and Knepper says that's not my imagination, but the work of receptors in the back of the throat that can influence perception. Water really does taste better when our bodies really need it, he says. "But if you're not dehydrated, it's not so good and may even be disgusting."

Shortly after the invasion of Iraq in 2003, Knepper got a call from doctors at Walter Reed Army Medical Center seeking advice about hyponatremia, which was causing a troubling number of evacuations. The military's standard operating procedure at the time called for troops to drink about a quart of water per hour while working in the arid environment of Iraq. Simply put, they were drinking too much, Knepper says. "The Army eventually cut back to half a quart per hour. I could not convince them to let the soldiers drink enough to abate their thirst." Water quotas are dangerous, he says, because they can lead people to overdrink and potentially end up with a swollen brain—or worse.

After examining the science, I can't help thinking we've made hydration unduly complicated. I take my dog running with me most of the time, and I've never measured the color of her pee or forced her to drink (as if I could). I make sure she has regular access to water, but she doesn't always take it. At times, she won't drink at all during a long run, and on those occasions she always goes straight to her water dish when we get home and slurps until she's satisfied. I've never had to give her an emergency IV for low fluid levels. If drinking to thirst is good enough for her, it's probably good enough for me too.

3

The Perfect Fuel

The first time I gave any consideration to postexercise refueling was in my high school physics class. During a unit on energy, our teacher, Mr. Gore, had us watch an episode of the PBS show *The Ring of Truth*, hosted by MIT physicist Philip Morrison.[1] To explain how the human body uses energy, Morrison turned to cyclists competing in the Tour de France and calculated the fuel they needed each day to complete the three-week stage race. He did this using what he called a "wholly commonplace unit"—the jelly doughnut. A normal person like him might do fine with the caloric equivalent of twelve jelly doughnuts per day, Morrison said. "But not for the racers!" he declared in his distinctive raspy voice. He then proceeded to pile doughnut after jelly doughnut onto a barbecue grill. "Thirty or 32 fine jelly doughnut units—that's the energy income each day of one of those athletes," Morrison said, before lighting them on fire to demonstrate how food is converted into energy.

"I eat a Hostess Twinkie before I get up," cyclist Alex Stieda of the 7-Eleven team said in the video.[2] After the race, Stieda ate "fruit—something that's easy on your stomach. Wait a couple hours, then we have dinner." To recover from their extended efforts, cyclists riding the Tour needed carbohydrates to replenish their muscle glycogen stores, protein to help repair muscle damage, and calories—lots of calories.

Morrison goes into the kitchen with a chef who's preparing the riders' meals and asks him to lay out on a table a full day's food. The spread includes coffee, pastries, corn flakes, bananas, peaches, ham sandwiches, bread rolls, butter, a green salad, soup, large cuts of red meat, multiple servings of some kind of tart, and a small glass of wine. "Look at it. It is familiar, attractive food. We all recognize it. Good French food, well prepared. That is not its special property. What is different is its quantity," Morrison said. As the story continues, we watch cyclists downing plates of pasta, pork chops, and sautéed vegetables, among other things.

The cyclists whom Morrison dropped in on in the 1980s fueled their engines with fine French food, but by the time I started bike racing in the 1990s, food specially engineered for athletes was becoming the norm. Around the time that Alex Stieda was starting his race mornings with a Twinkie, a couple of runners in northern California were experimenting with mixtures of syrups and grains in an attempt to create an ideal food for consuming with exercise. "We wanted to create something for ourselves," says Jennifer Maxwell, a nutritionist and runner who, with her husband, Brian (who died in 2004), started experimenting with various culinary concoctions. "When you're pushing beyond the typical weekend warrior and training at a high level, nutrition becomes paramount." At the time, she says, there was a growing emphasis on nutrition among athletes. The Maxwells developed their bars through trial and error, mixing up different ingredients on the old Wedgewood stove in their apartment kitchen. "We had certain parameters we wanted to meet. One was to have it be low-fat, primarily to make it digestible. Two, it had to have a shelf life," Maxwell says. As a base, they tried brown rice and some other grains before eventually settling on oat bran, which was then thought of more as cattle feed than a food item, Maxwell said. "We liked it because it was high in soluble fiber and would make sort of a gel." They tested their creations on other members of their small, tight-knit community of endurance athletes. "We would show up at weekend runs with bars wrapped in cellophane and tell people, here, try this and see how you feel," Maxwell said. Based on such feedback, the couple would go back to their kitchen and adjust the recipe a little more. One experimental version contained a milk protein that had digestibility issues, so they cut it from their array of possible ingredients. They played around with different ratios of simple to complex carbohydrates and various amino acids, vitamins, and minerals. They knew what they were going for as far as nutrients went, but they wanted their creation to be as palatable as it was nutritious. "The taste was paramount. It was sort of like a taffy." It took the Maxwells about three years to perfect the recipe, which also included some vitamins and minerals. They cautiously guarded their recipe, worried that someone would steal it.

They called their creation the PowerBar, and in 1987, they began marketing the product commercially. "It was our baby," Maxwell says. "We made it ourselves with our own hands." Those chewy bars in the gold wrappers heralded the beginning of a new era of sports nutrition. "We put rice flour on the outside of the bars, so they wouldn't stick to the wrapper," she says. "The TLC was very evident. We didn't cut costs. We touched every bar. When we started as a mail-order business, I would put little hand-written notes in the boxes." Athletes would write back to say they loved the taste and convenience of the bars. By 1989, PowerBar had acquired their own plant in Berkeley to produce the bars on a larger scale.

PowerBar sponsored my Boulder, Colorado–based cycling team in the early 1990s, and the bars that the company sent us by the case did indeed taste like taffy—rolled in sawdust. They came in two flavors—chocolate and malt nut. "The chocolate one was modeled after a Tootsie Roll—it was a healthy candy bar," Maxwell says. If you popped them in the microwave or a sweaty pocket and stretched your imagination a little, you could pretend that the chocolate ones were brownie batter mixed with wheat bran. In the winter, the bars could double as snow scrapers for your windshield, because they became hard as plastic in the cold. None of this bothered us, though. PowerBars were what every serious athlete ate, and we were thrilled and grateful to get them for free. Having their name on our jerseys was like a badge of legitimacy.

I had just graduated from college with a degree in biology, and I wanted to know everything I could about the science of sport. My cycling coach was working on a graduate degree and had spent a lot of time in the university's exercise physiology lab. She told me that one of the most important things I could do for my recovery was to eat a snack, preferably something with a bit of protein, right after training. She'd been following the most cutting-edge research, which was starting to show that not only did it matter *what* you ate after exercising, it also mattered *when*.

———

In 1998, a New Jersey company called PacificHealth Laboratories began touting a study that purported to show that their sports drink improved endurance performance by 55 percent compared to Gatorade. Not only that, the product was also said to improve recovery and might even protect against exercise-induced muscle damage. The study was presented at an American College of Sports Medicine meeting, and its inventors filed a patent for the product, dubbed Endurox R4.[3] While Gatorade had been sold as a beverage to quaff *during* exercise to maintain performance and hydration, Endurox R4 was presented as a new kind of replenishment—a beverage to boost recovery *after* exercise.

The drink was the brainchild of PacificHealth Laboratories founder Robert Portman and another sports scientist, John Ivy from the University of Texas. "I guess we were sort of the godfathers of recovery," says Portman, who holds a dozen patents on various nutritional interventions for health and athletic performance. Before then, people were aware of the importance of recovery, but no one had really focused specifically on the role that nutrients might play in the process, he says.

It was already well known that exercise depleted the muscle's energy, stored in the form of glycogen, and consuming carbohydrates after that exercise was important to help replenish it. But starting in the 1980s, Ivy's research had implied that this recovery process could be enhanced if carbohydrate was ingested immediately after exercise, instead of waiting until later.[4] "The rate of glycogen storage was twice as fast when you provided it immediately postexercise," Ivy says, because muscles become extra sensitive to insulin following exercise. As a result, they seemed to take up carbohydrate and store it as glycogen much more effectively immediately after a workout. That was Ivy's first insight. The second centered on the need for protein too.

At the time, carbohydrate was considered a fuel for runners and protein was thought to be what body builders needed. But Portman and Ivy suspected that repetitive aerobic exercise like running or cycling produces sufficient strain on the muscles to increase the need for protein too. "We thought that if you're dealing with muscle damage, then you're probably dealing with protein, and that was contrary to a lot of conventional thinking at the time, which said that you don't give protein to aerobic athletes," Portman says.

At least one study had suggested that taking protein supplements after exercise increased the rate of amino acid uptake and resulted in a faster rate of muscle protein synthesis. Giving exercisers protein immediately after a bout of intense exercise instead of waiting several hours seemed to speed its uptake in a manner similar to what they'd seen with carbohydrate. The lesson seemed clear: timing mattered.

Portman and Ivy contended that after exercise, there is an interval they called "the metabolic window of opportunity," when recovery could be accelerated by ingesting the correct nutrients. They dubbed this concept "nutrient timing," which was also the title of the book that the two men published in 2004.[5] "Nutrient Timing is not a commercial gimmick," they wrote. "Rather, it is the fruit of cutting-edge scientific insights into exercise metabolism, physiology, and nutrition." They divided their nutrient timing cycle into three phases—the energy phase, just before and during exercise; the anabolic phase, the time immediately after exercise and lasting up to 45 minutes; and the growth phase, when recovery and adaptations occur.

According to Portman and Ivy, taking in the right combination of nutrients during the anabolic phase could boost the rate of muscle glycogen storage, reduce muscle damage, raise the amount of protein taken up by muscle, and accelerate recovery. Portman was formerly an ad man, having previously founded two medical advertising and communications agencies, and he knew the power of marketing. Their book popularized the idea that my cycling coach had drilled into me—that ingesting nutrients at the right time was essential to promoting recovery.[6] Soon, athletes from a wide range of sports were rushing to refuel within the metabolic window, lest they miss out on a chance to enhance recovery and muscle adaptations.

With their nutrient timing hypothesis in hand, Portman and Ivy started looking for the ideal formula for a recovery product, and here they received some help from the late Edmund Burke, an exercise physiologist renowned for his work with elite cyclists. (He was also part of the medical team that gave performance-boosting blood transfusions, which were legal at the time, to US cyclists for the 1984 Olympics in Los Angeles.)[7] Looking at it conceptually, Portman decided that they probably needed antioxidants to counteract the metabolic damage caused by exercise. "You knew that protein synthesis was turned on, so you needed a substrate there, and you needed carbohydrates," Portman says. "We began playing with different formulas and evaluating them in a laboratory. The formula we arrived at was a ratio of carb to protein of 4:1—four grams of carbohydrate to one gram of protein, in conjunction with various antioxidants."

They named their creation Endurox R4 for the four Rs: restoring fluids, replenishing fuel, reducing muscle stress, and rebuilding muscle protein, a motto created by Burke.[8] Endurox R4 appears to have been the first drink marketed specifically for recovery, and its makers used science to make their case. Their studies put volunteers through some form of intense exercise to deplete muscle glycogen stores, then had them drink either Endurox or water. After a period of rest, participants would be exercise-tested again. "What we saw was an immense improvement in the subsequent exercise," after drinking Endurox, Portman says. PacificHealth claimed that Endurox R4's 4:1 ratio of carbohydrate to protein could significantly enhance hydration, improve endurance, reduce exercise-related muscle damage, and accelerate postexercise muscle recovery, compared to drinking a beverage with carbohydrate alone.

With the marketing of PowerBar, Endurox R4, and similar products, a new paradigm emerged—if you wanted to achieve peak performance, you needed a special nutritional formula, delivered at precisely the right time. Strength and endurance athletes alike all flocked to recovery drinks, protein powders, and specially engineered recovery foods. Gatorade, PowerBar, Clif, and other companies came out with their own protein and recovery-branded products, and vitamin shops created whole sections for protein recovery shakes. Protein powders and supplements had been popular among body builders for decades, in part because of the proliferation of articles and advertisements for these products in muscle magazines, which profited from their ads and sales. Today, you can walk into nearly any sporting goods store, specialty athletic shop, health food market, or supplement store and find an array of nutritional products marketed to promote recovery.

———

Portman and Ivy's book spread the notion of "nutrient timing," but the idea was popularized and commercialized before all the details were worked out. The problem with nutrient timing is that it suggests a sense of urgency and precision that has not held up in subsequent research. That carbs and protein are important for recovery is pretty certain, but the optimal amount and timing is less so.

In 2013, Brad Schoenfeld, director of the Human Performance Lab at CUNY Lehman College in the Bronx, and his colleagues published a meta-analysis looking into the evidence behind the idea of the "post-exercise anabolic window."[9] They concluded that the evidence didn't suggest the existence of a narrow window. The studies that supported the concept were essentially narrow snapshots in time that didn't capture the full picture of what happens over the course of a full recovery cycle. Nor were they necessarily generalizable to what athletes normally do. Part of the problem is that the original research looked at the response to a high dose of protein following exercise and compared it to a placebo. "That's not a proper design," Schoenfeld says. If your question is about the timing, you shouldn't compare a dose of protein given at one time to no protein given at all. Instead, you need to compare the same doses of protein given at different times. When Schoenfeld's team did that with a study looking at what happened if exercisers took protein right before versus right after a workout, they found no differences.[10]

There's surely a period where your body needs protein to repair and build after a muscle-straining workout, particularly something like a max session in the weight room, a CrossFit WOD (workout of the day), or a high-intensity interval session. But it's not so much an anabolic window, Schoenfeld says, "it's an anabolic barn door." As long as you eat breakfast, lunch, and dinner, it's almost impossible not to get through. The barn door doesn't slam shut 45 minutes after exercise. Instead, it stays open for four or five hours, maybe more. The latest research shows that protein will help recovery whether you consume it before or even during exercise. There's nothing magic about the 20, 30, or 60 minutes after a workout. The benefits come from the protein itself, according to Schoenfeld, not the exact timing of its consumption.[11]

How much protein do you need? For a while, studies seemed to suggest that the benefits of postexercise protein topped out at about 20 grams, but more recent research implies that some athletes with a lot of muscle mass may benefit from something more like 40 grams, says James Betts, a sports nutrition researcher at the University of Bath. What's the best number? Researchers are still debating the answer to that question, and the best way to resolve it is with more data.

According to what we know at the moment, smaller amounts of protein spread regularly over the day seems to be a better way of consuming protein than trying to eat or drink a bunch of protein immediately after working out. You don't need to take a protein supplement or shake to get your protein. "There's no evidence to show that protein shakes are any better than real foods," says Asker Jeukendrup, an Ironman triathlete and sports nutritionist who worked for Gatorade before launching his own sports nutrition business. Everyday foods are perfectly suitable, whether that means dairy products, plant sources like beans, or the little tins of tuna and salmon that have seen a resurgence in Australia.

But nutrient timing wasn't only about protein. Prolonged and/or intense exercise depletes glycogen stores in the muscles, and you need carbohydrates to restore them. Glycogen does seem to get replenished rapidly if you take in carbs immediately after a workout, says Schoenfeld. "But here's the rub. If you're not going to train again until the next day, there's zero relevance." Here again, the enticing initial studies weren't exactly wrong, but they were misleading. Their narrow scope and design mean they're not readily applicable to most real-life situations. What's more, the amazing benefits from immediate refueling appeared less spectacular in subsequent studies. Schoenfeld's analysis showed that as long as you take in some carbohydrates, your glycogen stores will be similarly replenished whether you consume those carbs 20 minutes after your workout or 3 or 6 hours later. On the other hand, if you're training again in a few hours, then by all means replenish your energy ASAP, but know that the reason to do so is that you're going to need that fuel again soon, not because waiting will impair your recovery.

As researchers have accumulated more evidence, it's begun to look as though the postexercise window just isn't as crucial as it was initially made out to be, says McMaster University protein metabolism researcher Stuart Phillips. "I don't really think it exists," he says of the magic window. "The idea that there was this time immediately after exercise when your muscle was made really sensitive to the provision of carbohydrate and your muscle would act like a sponge—a lot of that got dispelled by later research." Yes, it's important to get muscles the carbohydrates they need to replenish lost glycogen stores, he says. But whether that happens 30 minutes after exercise or six hours later, the results will be similar over the course of a day.

What makes the metabolic window of opportunity and recovery-branded products so appealing is that they promise to distill recovery to an exact scientific formula. But this precision is also what makes the claims dubious. The promises got ahead of the science. They're not entirely false; they're overstated. A finding that ingesting a particular drink or bar after exercise improved recovery became a claim that this product is essential and anything else is second-rate. This tendency to overstate is an ongoing problem in sports nutrition, says Louise Burke, head of sports nutrition at the Australian Institute of Sport. "I'm sure a lot of the people doing it aren't being malicious or deliberately misleading, but often when we try to market ourselves to athletes—even for the best purposes—we add a bit of pizazz." A new finding that looks and sounds hypothetically good and without risk becomes hyperbolized in the marketing, Burke says. What began as a cutting-edge idea that might still need some more fleshing out becomes a one-size-fits-all solution. But this cycle of hype glosses over the complexity of the science, Burke says. When she looks at that old Gatorade ad that advised athletes to drink 40 ounces of fluid per hour, she cringes. It's clear now that how much fluid an exerciser needs depends on a variety of factors, not a single formula set in stone. The same could be said of postexercise nutrient replenishment.

The crucial mistake happens when results of initial studies, which almost inevitably provide less certainty than the purveyors of these products imply, are taken as definitive. In fact, they're just the start. This is how science works—it's a process of slowly accumulating information and reducing uncertainties. First you figure out that protein is important, but then it requires more study to determine just how much and when. Science is a process of discovery much like the game 20 Questions, except that there are always more than questions and you can never answer more than one at a time—and that's if you're lucky.

———

There's a prevailing idea in sport that it's possible to perfectly optimize the body's physiology. Perhaps that's true, but our bodies are also extremely good at making do with what we throw at them. We're programmed to maintain homeostasis—a physiological state of equilibrium—even when conditions are less than optimal. Which means that it's important to get the big picture right, but fixating on the smallest details won't necessarily yield much payoff. The belief that there's some absolute perfect physiological state you can reach, if only you do everything right, opens the way to dubious products that use the language and jargon of science to exploit our search for the ideal.

We've been conditioned by a relentless tide of marketing to believe that we need specially formulated foods and beverages to refuel after exercise, but there's evidence to suggest that this idea is mostly a triumph of advertising. In 2015, a study appeared comparing how well cyclists recovered from a hard interval session after replenishing their energy with sports drinks and energy bars versus a calorically similar fast food meal.[12] Eleven male participants completed a 90-minute interval workout on an exercise bike at an intensity that would deplete muscle glycogen stores. Cyclists immediately recharged after this first workout with either Gatorade, Kit's Organic PB bars and Clif Shot Blocks, or hash browns, hotcakes, and orange juice from McDonald's. (The researchers chose McDonald's because it was conveniently located across from the school.) Two hours later, they had another snack consisting of the same category of food type, either Cytomax, PowerBar Energy Chews and PowerBar Recovery drink, or a McDonald's burger, fries, and a Coke. Four hours after the first workout, volunteers performed a 20-km time trial to test their performance. Researchers also measured muscle glycogen levels after the first ride and following the four-hour recovery period and took blood samples to analyze glucose, insulin, and blood lipids. A week later, participants returned to the lab and repeated the protocol, except this time, people who had fast food the first time got the energy bars and drinks, and vice versa. Overall, participants consumed a similar amount of protein, carbs, fat, and calories no matter which type of food they ate.

Like most sports science experiments, the study was small, but the results showed no differences between the two conditions. Whether they replenished their energy with designer sports bars or fast food, the cyclists' bodies responded the same way—glycogen stores went up by the same amount, their insulin and blood glucose levels responded similarly to the different meals, and their time trial performances were not much different between the two trials.

At the 2008 Olympics in Beijing, the Jamaican track team's coaching staff strictly advised their athletes against eating any food outside the Olympic Village because some of the unfamiliar delicacies served in local restaurants might upset their digestive systems. Dog meat, they warned, might end up on an athlete's plate if he ventured outside the village. World champion sprinter Usain Bolt was sure he didn't want to eat *that*, or anything that might jack his gut and hurt his chances at winning gold medals.

As "a Jamaican, I loved my jerk pork, rice, yam and dumplings. Sweet and sour chicken did not cut it for me," Bolt wrote in his autobiography, *Faster than Lightning*.[13] So he went looking elsewhere in the village. "There's an assumption that junk food isn't available in an Olympic complex, that we all eat super-healthy meals, but that couldn't have been further from the truth," Bolt wrote. After several days of struggling to eat unfamiliar food, he gave up. "Forget this," Bolt thought, "I'm getting some chicken nuggets." So he headed to McDonald's and grabbed a box of twenty chicken nuggets for lunch, and returned again at dinnertime for another round.

With 880 calories and 54 grams of fat, 48 grams of protein, and 52 grams of carbs, those twenty chicken nuggets were the protein and carb equivalent of more than two scoops of whey protein powder and 24 ounces of Gatorade, with a load of extra fat calories piled on. (The nuggets also contained about ten times as much sodium as two sports drinks would provide.) Bolt's teammates mostly just laughed at him, but hurdler Brigitte Foster-Hylton wasn't having it. "Usain, you cannot eat so many nuggets! Eat some vegetables, man. You're gonna make yourself ill," she said. He nibbled a few of the greens she gave him, but they tasted terrible. Out of frustration, Foster-Hylton handed him a packet of thousand island salad dressing, which finally rendered the salad palatable enough for Bolt to choke down. From then on, for every meal he ate some greens drenched with dressing along with chicken nuggets. "I devoured around 100 nuggets every 24 hours. I was there for 10 days, which meant that by the time the Games ended, I must have eaten around 1,000 chunks of chicken," Bolt writes.

Bolt wasn't just lounging around waiting for his three races. A sprinter's life during a big track meet is a series of intervals: race-recover-race. To reach the finals of the 100 and 200 meters, Bolt had to perform two preliminary rounds in each event, and he also had the 4x100 meter relay to run. Recovering between events would be essential to his performance. As it turned out, those nuggets of deep-fried chicken parts fueled performances that earned him three gold medals.[14] He would go on to replicate his three golds at the 2012 Olympics in London and then again at the 2016 Rio Games, where he was photographed chowing down on chicken nuggets once more.

McDonald's association with the Olympics endows it with a sheen of sportiness, but Bolt didn't win three gold medals in Beijing because he gobbled chicken nuggets. He won the medals because he's the fastest man alive. Those chicken nuggets were adequate, if not ideal, fuel to power him through his nine heats, and to help him recover his energy in between them. Feeling satiated and not worrying about gastrointestinal issues are surely worth a lot to an athlete preparing for his most important events of the season. Would Bolt have performed better eating some other recovery foods? Maybe. The better question is: How much difference would it make? Think of Bolt's body as a high-performance sports car. The fuel matters, sure, but it matters orders of magnitude less than the engine, and Bolt's is unmatched.

When your muscles are hungry for fuel, they don't care where the energy comes from, says Brent Ruby, the University of Montana scientist who supervised the cyclists study, done by his then-graduate student Michael Cramer. The fast food in this experiment may not represent the healthiest everyday food choices, but neither are those engineered recovery foods, which are highly processed and laden with additives, he says. Regardless, any of these foods work. "The muscle could care less. If you're dumping in carbohydrates, the muscle is going to be satisfied."

———

Even so, we're still left with the burning question: What should I eat after exercising? The easiest answer is: whatever your body is hungry for. In the NBA, for example, peanut butter and jelly sandwiches have become the go-to snack of choice. According to an article by Baxter Holmes in *ESPN Magazine*, "The Rockets make sure the PB&J is available in their kitchen at all times, in all varieties—white and wheat bread, toasted, untoasted, Smucker's strawberry and grape, Jif creamy and chunky— and offer 12 to 15 sandwiches pregame, with PB&J reinforcements provided at halftime and on postgame flights."[15] Holmes went on to describe how players for the Golden State Warriors became borderline mutinous in 2015 when a new head of physical performance and sports medicine, Lachlan Penfold, attempted to ban PB&Js from the team menu. When Stephen Curry complained about his missing PB&J (Smucker's strawberry, Skippy creamy), Penfold, who previously worked at the Australian Institute of Sport, told him, "Sorry mate. We're not doing sugar." Penfold no longer works for the team.

When asked whether the NBA had a policy on PB&Js, league commissioner Adam Silver told Holmes, "Our official stance is that it is a healthy snack." Leave it to the boss in the suit to be the voice of reason here. The idea that a few spoonfuls of jelly are going to harm a bunch of guys who just worked their asses off is almost as ridiculous as the assertion made by one nutrition expert quoted in Holmes's story that the pleasure attained by eating a PB&J is akin to that from taking heroin. With around 400 to 500 calories, 50 grams of carbohydrates, 20 grams of fat, and 10 grams of protein, a PB&J provides a pretty decent mix of nutrients and on game day, these guys' metabolisms are like furnaces.

———

What becomes clear when you look at the history of sports nutrition is that what's considered the perfect food to go with exercise is determined as much by culture and tradition as by science. When the Maxwells were formulating their first PowerBars, fat was the enemy, so everyone wanted low-fat foods. Today, Jennifer Maxwell says, those preferences have changed. "Natural" products and ones that appear less processed are favored over ones that contain long ingredient lists and additives. Our definitions of what constitutes "healthy" and desirable foods have shifted over time. Sometimes this happens in response to new research, but food fads and changing marketing trends contribute too.

Our beliefs and expectations heavily influence the way we experience different foods. Studies have shown that the very same wine tastes better when we think it's expensive than when it's poured out of a cheap bottle. And research also shows that people rate food as more delicious (and are willing to pay more for it) when they think it's organic.[16]

Consider the rebranding of chocolate milk. When I was a kid, chocolate milk was a special treat that my school only served on certain days of the week. But now the beverage has been repackaged as a recovery drink. In 2012, on the heels of several studies suggesting that chocolate milk could boost postexercise recovery, the nation's milk processors launched a "built with chocolate milk" campaign.[17] From a nutrient standpoint, it's a solid idea, says Betts, the sports nutrition researcher at the University of Bath. Low-fat chocolate milk contains protein and carbs as well as electrolytes like potassium and magnesium. "If we were to design a man-made supplement, it might look something like chocolate milk," he says.

In the words of an advertising firm hired to promote it, the milk industry's marketing blitz "sought to change the perception of chocolate milk—from a sweet, fattening beverage, to a healthy and effective option for refueling your body after exercise." The campaign cut a multiyear sponsorship deal with the

Ironman brand of triathlons that made chocolate milk available at the finish lines of select events and provided a means for promoting chocolate milk via athlete interviews and social media.

Part of chocolate milk's appeal is that it's a natural, readily available, and familiar drink, not some engineered product with a bunch of unpronounceable additives. A backlash against highly processed recovery products may also explain some of the appeal of coconut water (the watery liquid found inside a young coconut), a product marketed as "nature's recovery drink" based mostly on the fact that it's high in potassium, an electrolyte. It also has some carbs and sodium, but not as much as most conventional sports drinks. So far, studies have failed to find much evidence that coconut water is any more hydrating or helpful for recovery than other sports drinks. But appeals to people who yearn for "natural" foods seem to be working, judging by the proliferation of coconut water products.

Cherry juice has become another popular recovery drink in the "natural" category. A 2006 study found that people who drank juice from Montmorency tart cherries had less muscle soreness after a hard bout of arm exercises.[18] Since then, at least seven other studies have found some positive effect—improved muscle function and reduced inflammation are the most common benefits—from the juice, which can be taken as a liquid or in concentrate form. Montmorency cherries are a cultivar of tart cherry often grown in the Midwest, and they're particularly rich in antioxidants called anthocyanins and other flavonoids and also contain melatonin, and one study found that they could help with sleep.

The evidence on this cherry juice is promising, especially given that antioxidants are already known to reduce inflammation, but it's not yet definitive, and a recent review of fifty

studies on the use of antioxidants for preventing muscle soreness after exercise found that any benefits were too small to be
noticeable.[19] With cherry growers already sponsoring marketing campaigns, though, its popularity is likely to continue.

We seem to be suckers for "superfoods." Whether it's an old
standby like the PB&J, a rebranded staple like chocolate milk,
or a newly "discovered" item like cherry juice, we're all looking
for the magic bullet—the special ingredient or compound that
will give us an edge—and advertisers are standing by, ready to
provide it.

———

In today's high-tech world, many athletes want an app or scientific formula to tell them what to eat. But one of the most
fundamental skills athletes can develop is the ability to listen
to their bodies. And just as it does for hydration, the body has
some built-in intuition to help—hunger.

The problem is that our bodies' cries for food can be hard to
hear over all the advertising and cultural cues that constantly
bombard us. "A lot of people have lost touch with their ability
to pay attention to their body's hunger signals," says distance
runner Shalane Flanagan, a three-time Olympian, silver medalist in the 10,000 meters at the 2008 Beijing Olympics, and
2017 champion at the New York City Marathon. "I really try to
listen to my body."[20] Early in her career, Flanagan tried out bars
and other packaged sports foods, but found them unsavory.
"I wasn't excited to eat a bland bar that told me I could get 40
grams of protein, but it tasted like chalk." Elyse Kopecky, a college teammate of Flanagan's from the University of North Caro-
lina at Chapel Hill's running team, went on to become a chef
and nutrition coach after college, and the two friends published

a cookbook for runners, *Run Fast, Eat Slow*, that features simple foods made from mostly unprocessed ingredients. "Elyse
showed me that I could make my own recovery foods that are
really nourishing and taste good," Flanagan says. "Learning
how to cook for myself and eat good food was a game-changer."

Flanagan puts great care into how she fuels her body, but she
does it without a bunch of formulas or rigid rules. "I focus on
putting in the right foods, making sure they're whole foods,
then I don't have to stress about counting calories and worry
about whether I'm going to be light and lean enough. Just focusing on what you're actually putting in your body is the most
important part."

Many endurance athletes, particularly distance runners,
run into trouble by fixating on calories and trying to maintain a
light weight, to the detriment of recovery. "Light and lean is an
important part of the sport—the less you have to carry around
over 26 miles, the less effort you're going to have to put forth,"
Flanagan says. "But at the same time, if you're under-fueling
and under-nourishing the body, it won't be able to put forth the
effort it's capable of, because it doesn't have the energy and the
resources. The muscles start to break down and it's basically
cannibalism of your own body and that will lead to injuries and
other problems."

There's now compelling evidence that undereating can seriously impair recovery and adaptation and may lead to long-term
health problems. The issue was originally identified in female
athletes and called the athlete triad—named for insufficient
energy intake, which then leads to disruptions in the menstrual
cycle and impaired bone health. But researchers now prefer to
call the problem "relative energy deficiency in sport," or RED-S,
in recognition that male athletes are also prone to undereating,

and they also suffer metabolic, hormonal, and bone problems as a result.[21] Sure, staying light can aid performance, but it's a fine line. Eat too little and you become fragile and your performance and recovery suffer.

For Flanagan, the solution has been to focus on healthy foods, rather than fixating on calories. She stocks her kitchen and her cupboard with healthy items so she's sure to have good choices available. She often makes postworkout snacks, like "super-hero muffins" or "giddy-up energy bites," ahead of time, but in a pinch she'll grab a handful of nuts and dates.

———

For an athlete seeking to enhance short-term recovery—a swimmer or track runner performing multiple heats or events in one meet, for example—the focus, according to nutrition experts, should be to replace carbs and protein as soon as possible. "If you have only a few hours to recover before the next heat, then it becomes really critical," says Asker Jeukendrup, the triathlete and sports nutritionist formerly employed by Gatorade. On the other hand, if you've just run a marathon or played a weekly pickup soccer game and won't need to train or perform again right away, then timing is far less crucial.

Similarly, an athlete training in the preseason aims to maximize the body's adaptations to training and here it may make sense to forgo some short-term benefits in order to enhance long-term adaptations. "What is good for acute recovery can be bad for long-term recovery," says Jeukendrup. "Sometimes when you speed the acute recovery, you're actually harming the long-term adaptations." For instance, he says that antioxidants may help a little bit with short-term recovery, even though they seem to interfere with long-term adaptations. "There's some evidence that maybe the next day you're a little less sore, because you've suppressed some of the inflammation, but that inflammation is exactly what you need to adapt." Several studies have suggested that taking a high dose of antioxidants might blunt the training adaptation to exercise, although Jeukendrup cautions that the relationship between antioxidants and adaptation is still being worked out, and not all studies have found that antioxidants hinder adaptations (or even reduce inflammation).[22]

Some researchers are investigating whether athletes might sometimes want to hold off on replacing glycogen stores right away, with the idea that leaving the body under a bit of stress for a little while longer might actually enhance the adaptation you get from that exercise session. A similar concept, dubbed "training low," involves training without much carbohydrate available, either because muscle or liver glycogen stores previously have been depleted or because carbohydrates aren't consumed before or during exercise. The idea comes from observations that carbohydrate availability seemed to track with changes in the expression of genes thought to be involved in protein synthesis—an important process in the muscle's adaptation to training. Studies have shown that "training low" might produce beneficial molecular changes, but so far there's not much evidence that this translates into improved performance.

The train low, race high (with ample carbohydrate available) paradigm is intriguing, but without more evidence to understand how it works and what benefits it might have, it could easily become the next version of nutrient timing—an idea that started with some preliminary scientific findings, only to be expanded and oversold into advice that overpromises and underdelivers. It will take a lot more studies before the best ways to periodize nutrition are pinned down, but, for the

moment at least, the future of recovery nutrition appears to lie
in an approach that's periodized in much the same way that
most modern training plans are—focusing on promoting adaptation in the preseason and shifting to an emphasis on replenishment during the competition season. Whether this approach
will eventually go the way of nutrient timing or become lasting
gospel remains to be seen.

With all of the nutritional advice out there (much of it conflicting), perhaps the most important thing to understand is
that our bodies are not teeter-totters that require some very pre-
cise balance of inputs to avoid crashing down. Instead, we are
highly adaptable machines designed to maintain homeostasis—
our bodies' physiological equilibrium—by constantly adjusting
to our changing environment. The human body is remarkably
capable of recouping, even if we do everything wrong with our
recovery regimes, says Betts, the UK sports nutrition expert.
"If you don't have to come back and do another workout or competition for several days, nothing really matters. Your body will
get back to where it needs to be." Unless you have some need to
expedite recovery, like you're in the Tour de France and you have
to do another race the next day, the exact timing of your postexercise nutrition isn't do or die. "You can just go have a regular
meal and you'll be all right," Betts says.

The purveyors of sports nutrition products have every incen-
tive to promote the idea that you need their product, but the
notion that there's one single ideal food for recovery is not just
an oversimplification of the science, it can become counterproductive if ensuring that you get some specific food at an exact
time becomes its own source of stress. Whether you follow your
workout with a jelly doughnut, peanut butter and jelly sandwich, chocolate milk, or some designer juice, you'll be okay if

you're eating enough calories and nutrients and your overall diet is good. In a world where we're constantly encouraged to overthink what we eat, it can be a relief to let go of the obsession with attaining some absolute optimal diet and meal timing and instead trust your body to adapt to eating merely pretty well.

4

The Cold War

I n the photo, LeBron James gives a look that's a cross between a grimace and a snarl. The superstar forward then playing for the Los Angeles Lakers is submerged waist deep in a bathtub filled with ice cubes, and his powerful shoulders bulge as he holds his arms and hands up out of the frigid water. "Training camp ain't nothing nice! #MyFaceTellItAll #ThisTubColdAs YouKnowWhat #StriveForGreatness," reads the caption on the image, which James shared with his nearly 30 million Instagram followers. The message was clear: James wasn't just training hard; he was recovering hard too.[1]

Training camps like the one LeBron was attending that year can push athletes to utter exhaustion. This is by design—blocks of intense, multiple daily training sessions are intended to force athletes into a state of "supercompensation," where the body adapts to stress by fortifying its resources to become faster, fitter, stronger.

The road to supercompensation is paved in pain, and icing is a popular (if counterintuitive) way of coping with the hurt. Although it's painful at first, icing eventually numbs the

affected areas and people swear it reduces soreness. The practice has been around for several generations, but the icing age
really took hold in sporting circles around the time that physician Gabe Mirkin wrote about the familiar term RICE (Rest,
Ice, Compression, Elevation) in his 1978 tome, *The Sports Medicine Book*.[2] Mirkin didn't invent the acronym or the sequence of
actions, which had been mentioned in medical journals as early
as 1906, but he helped popularize it within the sports medicine
community.[3] For twenty-five years, Mirkin hosted a syndicated
call-in radio show about health and fitness, where he often mentioned RICE.

Intended as a way to speed the body's healing by shunting blood and inflammatory cells away from injured tissue,
RICE became a standard treatment for sprains and strains,
as well as for muscles that were sore from exertion. In the four
decades since Mirkin first promoted RICE for sports injuries, icing has become standard advice for people suffering
sprained joints, sore shoulders, and other orthopedic aches
or injuries. Who hasn't rolled an ankle and been told to put
an ice pack on it? Today, ice packs have become as ubiquitous
as aspirin—they're a fixture in every athletic training facility
and sold in drugstores throughout the country.

Cold baths and ice tubs have also become one of sport's most
popular recovery aids. Nearly every high school, college, and
pro trainer's room has at least one ice tub, and over the last ten
or fifteen years they've become an essential postworkout ritual
for athletes in every sport. One attraction is that icing is cheap
and easy. With a few bags of ice and a bathtub or even a garbage
can, any athlete can make a low-budget ice bath in a matter
of minutes.

Icing aficionados can be found in nearly every sport. Marathon world record holder Paula Radcliffe helped popularize cold treatments among runners by attributing some of her success to the ice baths that became her habit. Surfer Kelly Slater is also a fan. In an Instagram photo, he appeared sitting in a metal tub filled with ice water, his arms levitating stiffly above the water, as if he's on the verge of jumping out, and his lips puckered into an expression that screams, *ouch!* [4] During NFL training camps, it's typical to see whole teams of players shivering in tubs of ice. In the summer of 2016, an ESPN reporter tried some immersive journalism by interviewing Washington Redskins defensive lineman Chris Baker while the two sat in adjacent plastic tubs of ice water. And at one recent summer training session, ten Newcastle United soccer players were photographed crowded into a single inflatable kiddie pool filled with ice water. Even rock stars use ice baths. Madonna told *Rolling Stone* magazine that after her performances, which usually include hours of dancing, often in high-heeled shoes, she likes to take a ten-minute ice bath to recover.[5] "It's really painful when you get in, but it feels so good afterward," she said.

That sense of triumph is earned—an ice bath hurts. You can't ease into it, one toe at a time, because the natural reaction to this kind of cold is to recoil. Instead, you have to plunge in and submerge yourself before you have a chance to change your mind. The first few seconds you're fighting your body's impulse to get the hell out. After that, you're just marinating in steady anguish. Toes throb and sting. Feet ache. Large muscles, like your calves and quads, tingle and burn, and if you have genitals that dip into the icy water, they might try to rush back up the inguinal canal. "A man in an ice bath will never get an erection," said one expert who made me promise I wouldn't name him.

All this agony feeds into a culture of sport that idolizes grit and assumes that pain equates to gain. The fact that icing feels so excruciating almost surely adds to whatever effectiveness the technique might have. Scientists call this an active placebo effect—our natural inclination to believe that if a treatment is painful, it must be very effective. If it hurts, you assume it must be working, and this can influence your assessment of how much it helped.

The rationale behind recovery ice baths and cold tubs goes something like this: the cold stimulates your sympathetic nerve fibers, which react by signaling blood vessels in the area to constrict and send blood back to your core to protect your vital organs. This rush of blood away from the extremities reduces blood flow to the areas you're icing and slows the metabolic processes in these regions, including the inflammatory response, and thus reduces any swelling that might otherwise happen. The pressure of the water may also provide some compression against your muscles and blood vessels, which could also slow swelling and inflammation. Finally, icing relieves pain by numbing sore areas, at least temporarily.

That icing might suppress inflammation was originally a selling point. But in recent years, Mirkin, popularizer of the RICE method, has come to think of this as a bug, not a feature. In a complete turnabout, he now denounces the icing methods he once championed. There's no question that icing can reduce pain, at least temporarily, he told me, but it comes at a cost. "Anything that reduces your immune response will also delay muscle healing," Mirkin says. "The message is that the cytokines of inflammation are blocked by icing—that's been shown in several studies."[6] He now believes that instead of promoting healing and recovery, icing might actually impair it, and his change of heart came about largely because of one man.

———

Gary Reinl is a cold warrior. A slight man whose longish, graying hair is often topped with a baseball cap, Reinl has the fit, tanned look of a veteran marathoner. The Las Vegas–based personal trainer and entrepreneur has spent the past forty years working in the health and fitness industry. In that time, he's created rehab programs for injured workers and developed a strength-building protocol that's used in senior living facilities across the country. In the summer of 2016, he received a certificate of recognition commending him for outstanding achievement in support of the White House Medical Unit, and he's consulted for professional golfers, tennis players, CrossFitters, and NFL, MLB, NBA, and NHL teams. He's written books about post-pregnancy fitness and fat loss, but his most recent title, *Iced!*, is a self-published opus on the futility of icing sports injuries and tired muscles—a cause that has become his obsession.

Reinl is not a physician or a scientist or a coach. "I'm just a reporter trying to help people," he says. His journey to becoming an icing skeptic began in the summer of 1971. He was about to start his senior year of high school, and he was on a mission: to break his school's pull-up record. To succeed, he'd need to complete forty-two pull-ups from a dead hang. After some secret practice, he could do twenty-eight at a time, and so, confident that if he kept up his practice he could have the record beat by the end of the year, he announced to some friends that he intended to pursue the record. In the meantime, word of his plan got out and a rival announced that he could already do forty pull-ups and would try for the record too. "I was crushed," Reinl wrote in *Iced!* He considered the other guy tougher and more muscular than himself, and so he abandoned his goal. Later that year, the two boys tried out for the Marine Physical Fitness Team. One of the events was a pull-up test, and Reinl watched as his opponent struggled to complete fifteen pull-ups. "I was sick to my stomach," Reinl says. His blind faith in his rival's boast had broken his dream.

That, he says, was the last time he accepted an assertion without proof. "From that moment forward, I tried my best to do all that I could to identify and reject any unsubstantiated claims," he wrote. Since then, Reinl has made a career of questioning conventional wisdom. He has argued, for instance, that women could feel better during pregnancy if they do some strength training, and that older adults could abandon their walkers and wheelchairs if they take up strength training.

His latest mission is simple—to convince people to stop using ice on sprained joints and sore muscles. He likes to mention the Museum of Questionable Medical Devices in St. Paul, Minnesota.[7] "I won't rest until the museum adds another 'questionable medical device' to the curators' collection—the ice pack!" he says.

Over the years, he has worked with athletic teams, elite military squads, and coaches and trainers across the world. "The vast majority of people that I work with use ice—until they meet me," he says. "The first question I ask everyone is: why are you using ice? What are you trying to do? Most people say, 'to prevent inflammation,'" he says. "Well, why would you ever want to block or prevent inflammation? Without inflammation you won't heal! That's how your body regenerates!" His voice intensifies with impatience as he says this, struggling to conceal his frustration with the stupidity he's calling out. "When you ice, you don't block or prevent anything. All you do is slow things down so you don't heal as quickly."

In order to assess Reinl's argument, it's essential to understand what's happening in the body after an intense bout of exercise. One of the most common (and painful) aftereffects of a hard training session is muscle soreness, particularly the phenomenon called delayed-onset muscle soreness, or DOMS, that left me cobbled after my first Garfield Grumble running race. The agony of DOMS normally maxes out about 24 to 72 hours after the exercise, and it's most acute when you exercise your muscles in a way that they are under strain while they're lengthening. These opposing forces tug at the muscles, producing microscopic tears in the muscle fibers.

Your body responds to this injury by mobilizing a cleanup crew to remove damaged tissues and rebuild the muscles. This process fortifies the muscles, making them stronger. The repair response is also why a repeat bout of muscle-damaging exercise produces less DOMS than the first—because your muscle has become stronger and more resilient in response to the initial bout.

The cleanup and repair process is essentially the inflammation process, and Reinl's basic message is this: inflammation is your body's way of healing, and the only thing that icing does is delay this healing response. And that's true whether you're icing an injury or the micro damage from a hard workout. You can think of these insults as akin to freeway traffic accidents. When there's a wreck, you want the traffic flowing around it so that the ambulance and first responders can get to the crash site as quickly as possible. The same is true for an injured joint or sore muscle—you want the immune system to get to the site of the problem pronto, but icing shuts down blood flow and slows the immune system's response, Reinl says. It's like sending traffic off the highway and then closing the on-ramps so medics can't get to the accident. Instead of speeding recovery, this approach just delays emergency crews from doing their jobs.

Whether you're icing an injury to reduce swelling or cooling a sore muscle to tame inflammation, the approach won't work, Reinl says, because icing merely slows blood flow to the area, it doesn't halt it indefinitely. Once the icing stops and the blood flow returns to normal, whatever process you were trying to hinder will proceed again. The swelling will continue and the inflammation will start. The only thing you did was delay things. On this matter, Reinl managed to sway Mirkin, who wrote, in a foreword to *Iced!*, "Gary Reinl has done more than anyone else to show that cooling and immobilization delay recovery."

People are often shocked by Reinl's message, which in his animated telling can come across as a diatribe. But he's got research behind him. "People will say I'm an idiot, but not a single person has pointed out a mistake. And you know if they could find something, they would and it would be all over the internet," Reinl says.

Indeed, recent studies confirm Reinl's hunch that rather than speeding recovery, icing or cooling could actually hinder it. A 2006 study compared the training effects of a cycling ergometer or handgrip exercise with or without a cold bath afterward.[8] Participants did the same exercise on all limbs, but only one arm or leg was subjected to the ice bath. Over the course of four to six weeks of training, the cooled limbs made fewer performance improvements than their counterparts that were spared the cold plunge. Yes, ice can reduce the *pain* of swelling, but it doesn't seem to expedite the healing.

A 2013 study looked at what happened when cold packs were

applied to the exercised muscles for 15 minutes after a bout of arm extensions.[9] It turned out that subjects who iced had more fatigue than those who didn't, and the cold packs actually delayed recovery. Similarly, a 2015 study reported on two experiments looking at how cold water immersion influenced how muscles responded to a strength training program, and found that cold treatment reduced gains in muscle mass and strength and blunted the activation of key proteins in the skeletal muscle.[10] The studies "challenge the notion that cold water immersion improves recovery after exercise," the authors wrote.

With evidence like this backing him, Reinl has begun to gain traction among some leaders in the sporting community, including Kelly Starrett, a well-known physical therapist and founder of San Francisco CrossFit and Mobility WOD.[11] Starrett posted a 26-minute video of himself listening to Reinl's anti-icing pitch, and the clip has been viewed more than 200,000 times on YouTube.[12] "Ready to slay a sacred cow?" Starrett wrote in a companion blog post.[13] "You should stop icing. We were wrong." Starrett told me that, like many physical therapists, he'd once been a proponent of icing, and it had been a difficult belief to overcome. "I personally had a difficult time slaying this icing error beast until I was confronted directly with the physiology," he wrote in a message to his Mobility WOD members. "It's hard to fight the drag of orthodoxy. Why do we do what we do? Because we always have? We can do better." He urged his followers to "kick the ice habit."

Reinl's movement, which he calls "the meltdown," comes at a time when icing seems as popular as ever, with new cooling devices, such as ice sleeves and cuffs, being relentlessly promoted on social media.[14] Even in light of the evidence that icing and cold therapy could have downsides, not everyone is ready to leave icing behind, at least not entirely.

———

Shona Halson is the world's leading expert on recovery.[15] A recovery physiologist and former Head of Recovery at the Australian Institute of Sport, Halson has a warm, cheerful manner, and her Australian accent gives her speech an air of friendliness. She was raised in a sporting family—her father was a P.E. teacher and football coach—and she grew up running track and field. "I liked sprinting as fast as I could. One hundred meters was too far, 60 meters was about right," she says. She also played tennis competitively, but her performance was "never very earth-shattering," Halson told me. "I am quite typical of a lot of [sport] scientists. I competed a little, but wasn't good enough to make it to the big league." She likes that working with athletes helps her stay connected to sport.

Halson got an honor's degree studying chronic fatigue syndrome, and her PhD work focused on overtraining in athletes. "My interest in recovery came from my interest in fatigue," she says. But fatigue is a complex phenomenon that's difficult to study, so after her doctorate she turned to studying recovery, because it's (somewhat) easier to measure and assess. It was also an area of growing interest in the research community and among athletes who recognized that as they increased and optimized their training, they needed to master recovery too. Halson has been at AIS since 2002, and in that time she's worked with Olympians from a wide variety of sports. "Swimming and cycling are the two that I spend most of my time with," she says, but with surfing due to make its debut at the 2020 Olympics, she's just started working with the Australian surf team as well.

Halson knows the recovery literature inside and out, having been a leading contributor to the body of research, but she also has more than fifteen years of experience applying the science to athletes. Her perspective is informed from both angles. There are basically two competing theories regarding icing and cold therapy, she says. The first is that by decreasing inflammation, ice baths might stunt the body's ability to adapt to whatever training has just occurred. The other is that if an ice bath can reduce pain and soreness, then the athlete could potentially train harder, sooner. Which theory is correct is still up in the air, and it might be that they're both right, depending on the circumstance, she says. The research on this issue is still rapidly evolving, and more studies are needed to make the answers more definitive.

What researchers are beginning to realize, Halson says, is that recovery may be most effective when it's done in a periodized manner, just as training is. The idea is that rather than using the same recovery techniques day in and day out, you match the recovery to the goals of that particular training. When athletes are in the preseason part of the training cycle, the goal is to force adaptations that will boost performance in competitions later on. In that case, adaptation is the most important thing, and it's worth giving up a little bit of performance in the short term if you can make bigger gains in the long run. This is an argument against those ice baths that are so ubiquitous during the NFL and NBA preseason training camps. On the other hand, Halson says, when an athlete is in the competition season, most of the adaptations have already been achieved, and the goal is short-term performance. The type of training and its purpose should drive what kind of recovery methods an athlete uses.

Given the current evidence, Halson says, ice baths are probably not a great idea after strength training, if your goal is to get stronger. They're also not the best approach if the athlete is in a building phase of training where the goal is supercompensation. In light of what's currently known, the situations where athletes are most likely to benefit from icing or cold water immersion are ones in which they are seeking short-term recovery in between events and are not worried about long-term adaptations. In other words, she says, if it makes you feel better, go ahead and ice between prelims and finals of your track meet or between events at the CrossFit Games. She also sees a role for cold therapy when an athlete is feeling excessively tired. But if you want to maximize your results from training camp or a hard workout, she believes it may be best to skip the ice.

In practice, Halson has found that athletes generally report small, but meaningful improvements in how they feel after cold water immersion. A 2011 meta-analysis published online in the *British Journal of Sports Medicine* also estimated that cold tubs reduced perceived soreness by an average of 16 percent. But the study also identified a major flaw in the studies that weakened their conclusion.[16] The problem is that there's no easy way to blind participants to whether or not they're getting the treatment—if you get an ice bath you'll know it. And if volunteers know they're getting something that's supposed to help their recovery, they're susceptible to the placebo effect. The expectation that they will benefit may nudge them to perceive an improvement.

Is a 16 percent reduction in soreness a meaningful difference? Maybe. Previous studies have calculated that for a decrease in pain to make a real difference in everyday life, it needs to be on the order of 14 to 25 percent.[17] That implies that reductions in muscle soreness seen in these studies were just on the border of making a noticeable difference in anyone's life.

It may turn out that an ice bath's greatest benefits are psychological. A 2014 study employed a clever placebo group to test the effects of a cold water tub. Researchers had the placebo group apply a fake pain-relieving cream and then soak in a room-temperature tub, while a control group just took a dip in the pleasant water.[18] It turned out the placebo was just as effective as the cold soak. The authors note that muscle soreness is subjective. And if that's the case, does icing help soothe muscle aches because it's altering some physiological process, or is it just a matter of people expecting less soreness so that's what they report? It's hard to say.

More and more people are beginning to question the magic of icing. Malachy McHugh is director of research at the Nicholas Institute of Sports Medicine and Athletic Trauma in New York City. A sporty Irishman, McHugh plays Gaelic football (an Irish game often compared to rugby), and his team takes turns on the same playing field with the Manhattan College soccer team. One evening when the other team was finishing up, a group of soccer players were lowering themselves into trash cans filled with ice water as McHugh's team was taking the field. Seeing McHugh approach, they called out to him while pointing to their makeshift ice baths. "Mal—you're the sports scientist. What's the deal with this? Does it work?" His reply? "Eh, you'd be better off putting your six pack of beer in there."

I can't help thinking that what cold soaks do best is give people a recovery ritual and a sense of agency—the feeling that they've done something good for themselves. Surely that's worth something, even if it doesn't change your physiology. If, like marathon record-holder Paula Radcliffe, you make a habit of ending your hard efforts with an ice bath, the cold water might also trigger a potent psychological response. In the same way that dessert signals that the meal's over, an ice bath could tell your body the workout is done and it's time to wind down. And frankly, there are more pleasant ways to get that.

————

He may have won a few battles, but Gary Reinl is a long way from winning the cold war. Witness the growing number of athletes taking ice baths to another level with cryosaunas and cryochambers, which use refrigerated air or liquid nitrogen to quickly expose the body to temperatures as low as −250°F, a technique often referred to as "whole body cryotherapy." The practice originated in Japan in the late 1970s with Toshima Yamauchi, a physician who specialized in treating rheumatoid arthritis. Yamauchi noticed that his patients seemed to improve after going on winter holidays, especially when they went skiing or exercised in cold. He invented whole body cryotherapy in an attempt to replicate this approach in the clinic, and he owns several US patents related to cryotherapy.[19] The first cryochamber was made in about 1980, and the technique soon spread to Poland and Germany.

Whole body cryotherapy began as a treatment for rheumatoid arthritis and other inflammatory disorders, but it eventually made its way to the world of sport. The Dallas Mavericks used cryotherapy during their 2011 NBA championship season, when point guard Jason Kidd described the technique as a secret weapon for rejuvenation. The Kansas City Royals baseball team, Nike's Oregon Project running program, the University of Missouri athletics department, and the Dallas Cowboys football team are among dozens of other elite sports programs that have used the technique.

Cryotherapy is "like an ice bath on steroids, but without the searing pain or discomfort," says chiropractor Ryan Tuchscherer, who runs a series of clinics around the Denver region where anyone willing to pay about $50 can spend two to three minutes with nitrogen-cooled air swirling around them. His clients include LA Kings goalie Jeff Zatkoff, MMA (mixed martial arts) fighter J. J. Aldrich, and numerous members of the Denver Broncos, including linebacker Von Miller and cornerback Aqib Talib (now with the Los Angeles Rams).

"I use cryo when my body is really sore. It helps me recover and gets me back to where I need to be on the field," says Talib, who has played in the NFL since 2008. Earlier in his career, he says, he paid less attention to recovery. "Then I realized that the guys who play eight, nine or ten years, they really take care of their bodies." Now he views recovery as an important part of the job. "You've got to treat your muscles like it's your tools, it's your legs, it's your wheels. You have to take care of them." He also uses ice baths and contrast baths (a variation on cold water immersion where the athlete alternates between cold and hot tubs in order to promote circulation), but these take a lot more time. With cryotherapy, he only has to stand in the chill for three minutes. "Once you get about two minutes in, it's going to start feeling really cold. It hurts like hell, but you only have to go through that for one minute in the cryo, as opposed to in the cold tub, where you have that for maybe five minutes before you get numb," he says. "I use it a lot during training camp when we're really working hard and my body has to recover fast for the next day's practice. Me and Von Miller will leave training camp and go right to the cryo."

Not content to just take Talib's word for it, I visited Tuchscherer's 5280 Cryo & Recovery Clinic in Denver on a bitterly cold December day, the kind of winter morning where it seemed foolish to step out of the house, much less into a frigid vat of cold gas. But I was a reporter on a mission—to try out this high-tech thing I'd been hearing so much about. I'd gone for a long run the day before, and was eager to find out if a cryosauna could take some of the soreness out of my legs.

Tuchscherer has brown hair, cut short, a well-trimmed beard, and crystal blue eyes. Dressed in jeans, a Broncos hoodie, and a baseball cap with his clinic's logo, he met me at his clinic looking as much like a friendly guy at a sports bar as the chiropractor-entrepreneur in charge of a vast recovery center enterprise. Tuchscherer developed his financial acumen watching his father run a chain of pharmacies in Texas and Oklahoma, and he's involved in multiple businesses—clinics in Colorado and Kansas and new facilities in the works in several other states. In addition to the clinics, he sells Angus beef from cattle he runs on his family ranches in Kansas and North Dakota.

Located in an upscale office park, his Denver clinic was bright and cheery, with hardwood floors and lots of light. The clinic's "man in motion" logo, a clever schemata of a sprinter with gears for joints, was prominently displayed on a bright blue wall behind the front desk, as well as on the cryotherapy machine itself—to ensure that any selfies posted to social media would bear the proper branding, he explained.

The cryochamber itself was positioned in a small room that you might find at a doctor's office. A closet in an adjacent room held the liquid nitrogen for cooling the chamber, which was stored in a metal tank approximately the size of an NFL line-backer. A small gas line connected the tank to the cryocham-ber through a hole in the wall. Tuchscherer told me that these devices cost between $50,000 and $60,000 a pop. His clinics have ten of them, including two mobile units that they take to gyms and events in an RV. In January 2018, Tuchscherer launched another cryotherapy and recovery facility at a football training facility opened in Dallas by Aqib Talib and Von Miller.

To prepare for my session, I stepped into a private changing room and stripped down to nothing but the robe, dry socks, and fuzzy gloves that Tuchscherer had given me. He warned me against wearing my own socks if they were even the slightest bit damp—any liquid water would immediately freeze against my skin. Former Cleveland Cavaliers guard Manny Harris was reported to have suffered severe freezer burn on his foot in 2011 after entering a cryotherapy machine with wet socks.[20] Track sprinter Justin Gatlin similarly developed frostbite on his feet before the 2011 world championships after undergoing cryotherapy with sweaty socks.[21]

Dry and totally naked under the robe, I was ready to chill. Tuchscherer showed me to the cryochamber. It looked like a big steel drum. The front of the device swung open like a refrigerator door and the interior was lined with blue foam padding. I took a deep breath and climbed inside. Tuchscherer adjusted the height of the floor padding so that only my head stuck out of the cylinder, then closed the door.

"Okay, whenever you're ready, you can hand me your robe," Tuchscherer told me. I wasn't expecting that, and I felt chilly just thinking about what was to come. But I was here for the full experience, so I took off the robe and handed it to him over the top of the tank. Now I was nude in a room with this friendly stranger who was about to blast me with frigid air. As strange as it seemed in the moment, I reminded myself that this was a privilege that people paid a lot of money for. Tuchscherer asked if I was ready, and I gave a thumbs-up. And just like that, he started the flow of nitrogen. It felt like standing naked in a gentle windstorm on a winter day in the negative digits, and the stream of gas felt noticeably colder at my feet and lower legs than it did as it climbed up the rest of my body. I wore a ski hat throughout it all, and held my gloved hands up above the tank, which undoubtedly helped me retain some body warmth. A number pad on the top of the tank displayed the purported temperature in the tank as it fell—11°F, 7°F, 1°F, –2°F, –18°F, –21°F, –50°F, –100°F, –133°F, –148°F, down to –210°F and below. The temperature continued dropping the entire two-and-a-half minutes I was in there.

About a minute and a half in, Tuchscherer told me to rotate in a slow circle, mostly to distract me from the cold, he admitted later. The treatment was complete before I could have finished singing "Cold as Ice," but I could hardly sing as my lips puckered and my warm breath created whiffs of white fog. I wasn't quite shivering, but I was close as I breathed in staccato beats, thinking the whole while, *cold, cold, cold, cold, cold.* I was just getting acclimated when, just like that, it was over. For a moment, I felt blindsided—what just happened? It took a minute to process.

Tuchscherer handed me the robe, and I felt an amazing adrenaline rush as I stepped out of the chamber. My legs felt numb but invigorated. I'll admit, I was skeptical going in, but afterward, I totally understood the appeal. It was easy to convince myself that something profound had just happened. There was virtually no pain, only a pleasing coolness in my leg muscles. Sure, the treatment was cold, but it also produced a very pleasant rush that had an addictive quality to it. I would eagerly do it again. I got out of the tank feeling energized and slightly amped. I totally understood why so many MMA fighters liked to do this before a bout. Enduring the blast of cold made me feel powerful. I was ready to kick some ass. As far as experiences go, I was sold.

As the nitrogen had flowed and I stood there naked and locked inside the frigid chamber, Tuchscherer waved his hands around explaining, in animated terms, the rationale behind the treatment. The problem with ice baths, he told me, is that blood flow to the limbs is never 100 percent shut down. "The heart is still trying to push blood out to the extremities. With the cryochamber, we get the chamber inside cooled down so quickly that in a two-and-a-half- to three-minute period of time, *all* of the blood is pulled out of the tissues." From the tissues, he said, the blood is pulled into the core of the body, primarily the heart, and "the heart spins it, superenriches, superoxygenates it. As you get out, that superenriched, re-oxygenated blood has to go somewhere so it starts pushing back out to the body for the next six to eight hours," he said. "It's kind of a legal way of blood doping." The treatment only lasts a few minutes, but its effects last much longer, he said.

These claims sounded amazing, but I wanted an expert opinion and some science to back them up. Darryn Willoughby is an exercise scientist at Baylor University who wrote a review of the sparse research on cryotherapy. When I told him about the claim that cryotherapy could "superoxygenate" the blood, he chuckled. "Oh my lord! Are you serious?" He said that it's physiologically impossible to superoxygenate the blood, because the blood leaving the lung is already nearly 100 percent oxygenated under normal conditions. "Our blood is only capable of carrying so much oxygen. There's no such thing as superoxygenation," he said. The idea that cold exposure pushes blood to the central part of the circulation system is correct, but this all happens very quickly and also resolves quickly. Shunting blood to your core doesn't inject it with extra oxygen. So much for the legal blood doping.

Willoughby's conclusion? "This is just another fad. It will go away in a few years, like a number of other things. They always do," he said.

Tuchscherer also claimed that cryotherapy stimulates a fight-or-flight response that provokes the release of endorphins and increases the body's natural anti-inflammatory molecules. This claim seemed slightly more plausible. I definitely felt the kind of fight-or-flight rush he described. It was the same feeling I've had after recovering my balance mid-fall while skiing, or after having a close call while driving. But whether any of this results in improved recovery remains unproven.

Cryotherapy enthusiasts suggest it can superchill the body, but I was dubious. The gas may have been $-210°C$, but surely it would have hurt a lot more than it did if my skin got that cold. It also didn't seem plausible that three minutes in the cryo-chamber would cool my muscles as much as 15 minutes in an ice bath, for the simple reason that water is a better conductor of heat than air is. Basic physics, as my dad would say. Indeed, one study of whole body cryotherapy found that even with temperatures of $-180°C$ measured at the outlet of the device, the decreases in temperature recorded on the skin were still only between $-4°C$ and $-14°C$.[22] The study found that muscles cooled even less, about $-1.1°C$, and concluded that these differences are smaller than what you'd experience by using an ice pack or cold tub. In other words, there's a reason that cryotherapy is more pleasant than an ice tub—despite the low temperatures, you're not actually getting as cold.

In 2016, the FDA put out a consumer warning that there is

very little evidence about cryotherapy's supposed benefits or its effectiveness in treating the conditions for which it is being promoted, which include not just exercise recovery, but also arthritis, multiple sclerosis, fibromyalgia, Alzheimer's, and chronic pain. The FDA also made clear that the safety of using cryotherapy has not yet been established.[23] A 2015 Cochrane review examined the science on whole body cryotherapy and found that the existing studies were all of low quality, because there's no convincing placebo, and some of the claimed benefits, like soreness, lack an objective measure and are susceptible to the placebo effect.[24] The review concluded that there's insufficient evidence that it helps or that it's safe. More than twenty studies have been published on cryotherapy, but none of them offers convincing answers.

Tuchscherer emphasizes that the cryotherapy his clinics offer is a consumer product, not a medical device, and he makes no claims about FDA approval. Instead, a clinic brochure points to endorsements "by many doctors, professional athletes, sports teams and trainers," and its appearance on television shows like *The Dr. Oz Show* and *The Doctors*. But these shows are hardly known for their rigorous scientific analysis.

Cryotherapy definitely made me feel good. But did that momentary boost provide any lasting benefits? I stayed around for about an hour afterward, talking with Tuchscherer about his clinic and his other practices. He also provides chiropractic care, and he demonstrated an "Active Release Technique," a kind of physical therapy–like movement, on my sore hamstring. Afterward, Tuchscherer asked me if my muscle felt any better. I couldn't really detect any noticeable difference, but that kind of achiness is hard to quantify. I mean, the more I thought about it, the more I found myself thinking, well, maybe it *did* feel a tiny bit better, but it was hard to say for sure.

He'd been exceedingly generous and nice, and the thought of telling him that his special trick hadn't done anything useful made me feel unkind, so I said, sure, maybe it did help. Afterward I wondered, did the people in those cryotherapy studies feel a similar obligation to improve? I'd felt great immediately after the cryotherapy session, but by the time I left, I didn't feel much different than when I'd arrived. The cold rush was nice, but I could have gotten a similar effect by stripping down and making a snow angel in the fresh powder, for a lot less money and effort.

5

Flushing the Blood

The theory behind training and recovery boils down to this: you stress your body, and it responds by fortifying its resources to better handle the stress. You put strain on your muscle by lifting a weight, and it fixes this minor damage by reinforcing the muscle fiber so it's stronger for next time. How quickly you rebound from the strain of training by building strength and endurance depends on the amount of stress you experience and the resources you have available to work with. Imagine that your body is a house, and training and other stressors are the weather and elements. You begin with a house made of straw, and your first bout of training is like a gust of wind. It knocks out a few walls and so you build them back up. If you have the means, you'll probably build the new walls from brick. When the next bout of exercise comes along, your walls are more resilient, and this time nothing crumbles. You keep training, and as you do, you increase your training load, or stress, by lifting more weights, running more miles, or throwing more pitches. This bout of training is a storm, which perhaps breaks a few windows or blows off some roof tiles. Again, you get out there and fix the damage, and if your resources aren't already tied up elsewhere, you repair the damage so that the house can withstand even more strain next time. Maybe you replace the single-paned windows with double-paned ones or replace old wood shingles with fire-resistant ones. This storm/repair/repeat cycle is like the training cycle, and the way you get fitter, stronger, and faster is by subjecting your body to lots of stress, so that it "supercompensates."

The new focus on recovery recognizes that training is only as good as the recovery that follows it. If you think about the house and storm analogy, what determines how fit you'll get or how strong your house will become isn't just the degree of bad weather it's exposed to (i.e., the intensity of the training), but the quality of the repairs you make (i.e., the quality of the recovery you undertake) in the breaks between storms. And that's where recovery "modalities" swoop in to help.

Every recovery trick or product is a little different, but a vast number of them aim to boost blood flow. That's with good reason. Blood is the body's great delivery system. It helps get recovery done. It's what shuttles the metabolic by-products of exercise away from the tissues and brings the inflammatory molecules that help with repairs to the sites that need fixing. Blood also delivers oxygen to cells and glycogen to depleted muscles. The circulatory system is like a big city's transportation network—it allows things to get from one part of the body to another. The better that traffic flows, the more efficiently the body's work can get done. Protein can get to the muscles to rebuild damage, glycogen can get to the muscles that have depleted this energy source, and waste products can be sent to the kidneys or liver for processing and removal. Given all this, it's no wonder that virtually every recovery gadget and method makes some claim about increasing blood flow. Not all of the reasons for promoting circulation stand up, though.

When I was a high school runner in the late 1980s, my coach would tell us to shake out our legs to "get the lactic acid out." Back then, we thought lactic acid was what made our muscles sore. But that thinking was wrong, says Michael Joyner, a Mayo Clinic exercise scientist. Lactic acid, which is made up of lactate and an acid molecule, is produced in your muscles during heavy exercise, but it turns out that it's probably not responsible for the burn I felt in my legs at the end of a hard 400-meter interval on the track. Nor does lactic acid make muscles sore following exercise. In fact, research by George Brooks of the University of California–Berkeley showed that lactate may actually provide a source of fuel for muscles in some cases and may even help trigger the production of new mitochondria, the structures in cells that produce energy. While it's still true that exercising hard increases lactate levels and lactate buildup appears related to fatigue, the idea that lactate needs to be flushed from the muscles after a workout is misguided, Joyner says. "Lactic acid is removed from muscle very quickly during active recovery. It's gone in less than an hour. Even if you do nothing it will go down fairly fast." Increasing your blood flow only speeds things a wee bit.

Still, circulation is an important factor for expediting the cellular processes involved in recovery, and if you want your blood to flow, heat can help. Not only does heat cause your vessels to dilate, increasing blood flow, it usually feels pretty good too. I'm a fan of warmth—where an ice bath makes me feel stiff and numb, a hot shower or soak in a hot spring or tub always leaves my muscles feeling relaxed and loose. So I was probably predisposed to fall for a heat-related buzzword that I kept seeing again and again: "infrared."

———

The first thing I noticed about the infrared sauna was that it wasn't very hot. That was okay with me. It was an August afternoon, and I'd just spent three hours mountain biking up and down the steep slopes around Crested Butte. Heat wasn't the first thing that I yearned for just then, but I'd already made an appointment to try out this recovery method I'd been hearing so much about. It seemed like every recovery center I'd encountered had an infrared sauna, and spending time in them was increasingly touted as an essential recovery tool.

So I dropped in to Sea Level Spa and paid thirty bucks to spend 30 minutes in an infrared sauna. The place had a bit of a hippie vibe, and when I arrived a couple of middle-aged tourists were lounging in the front room, breathing in supplementary oxygen through nasal tubes and talking about how thin the air was up here in the Rockies. In addition to oxygen, the spa also offered a makeshift hyperbaric chamber, a wooden contraption that looked as if it had been made in high school shop class. For a fee, you could crawl inside the vessel, which was pressurized to mimic the air at sea level.

These tools for lowlanders were all well and good, but I've lived at altitude for most of my adult life and had come here for the special sauna. "It's more about being in front of the element than about the heat," the bearded guy at the front desk told me as he showed me to my private sauna and demonstrated how to open a little window in the door if I got too hot. It looked like an ordinary wood sauna. The only difference I could see was that instead of one large heater or bed of coals, it had four small heating elements that looked a lot like those electric heaters that restaurants set out on their patios to keep diners warm. The heaters made a sound like a gentle fan.

I got undressed and stepped inside. The sauna was about the size of a compact car, but there was enough room for me to stretch out my legs on the wooden slats. I leaned back and waited. Some speakers above me emitted soft, flute-heavy versions of "Ode to Joy," "Angel of the Morning," and "Greensleeves." According to the digital readout, the temperature started out in the upper 90s (°F) and topped out at 117°F (a traditional sauna is usually set between 150° and 220°). Perhaps because I wasn't feeling particularly cold when I got in, the warmth felt agreeable but not especially relaxing. I closed my eyes to concentrate on the experience in all its dimensions. I wanted to sense what made this sauna special, but whatever it was, it evaded my best efforts to detect it. I expected to feel something unique or enticing or, at the very least, ultrarelaxing. But I just felt that I was sitting naked on a warm wooden bench. The music gave the experience a New Age sort of feel, but my muscles didn't seem to care. I felt dry and sort of warm. That was it.

After the temperature gauge had read 117°F for a while, I opened the window in the door to let in some cooler air, and the little breeze felt quite soothing. After about 15 minutes, I was bored and ready to leave, but I felt obligated to stay for the whole experience. My anxious waiting did not feel relaxing. There must be some reason they set sessions at 30 minutes, but I never found out what it was. The last five minutes seemed to stretch on forever. I wasn't feeling flush like I do in a regular sauna. But I was starting to feel a little bit of sweat coming on, and I was ready to cool off. I'm pretty sure I would have enjoyed the experience more in the winter when it was cold outside. Going from a warm day to a warm sauna felt like a weird thing to do. Simply sitting outside in the warm Colorado sun probably would have felt just as good, if not better, but maybe the infrared benefits hadn't hit me yet. I wanted to keep an open mind.

So I called up a couple of infrared sauna makers to find out what I'd missed. What makes an infrared sauna different from a traditional sauna is the wavelength of radiation it uses, said Raleigh Duncan, founder of Clearlight Infrared Saunas. A quick physics lesson: infrared radiation is a type of thermal radiation that lies just to the right of visible light on the electromagnetic spectrum, so it has a lower frequency and higher wavelength than light that we can see with our eyes. Near infrared is closest to visible light, and far infrared is further down the spectrum. Near infrared waves are invisible to us—they're the kind of frequency used by your TV remote control. We sense far infrared as heat.

So—to translate: what makes an infrared sauna different from a regular sauna is that it uses heat radiation. Wait, what? Isn't a normal sauna also radiating heat? Yes, but it uses a different *kind* of heat. What we experience as heat happens when energy enters our bodies by jiggling the molecules inside us. This can happen in a variety of ways. A regular sauna uses convective heat—hot air rises from the heating element and hits us with energetic molecules. These molecules slam into our skin, and the skin transfers this energy to our blood and the rest of our body. An infrared sauna uses radiative heat—photons of a particular wavelength radiate from the heater and are then absorbed by our bodies.

Why does that matter? Duncan told me that traditional saunas heat up the air, whereas the infrared sauna "penetrates the skin, into the soft tissue," heating the body from within. "Initially it doesn't feel like much is going on, but then you start to absorb the infrared heat even though the air temperature hasn't changed," he said. "Essentially that's the difference." In both kinds of saunas, something is jiggling your atoms and molecules and making you feel warm. The difference is just in how that jiggling is instigated. It's like the difference between heating up your food in a traditional oven versus a microwave oven. The sauna is just a very inefficient microwave for your body.

If the infrared sauna I tried had only heated me from the inside, it sure fooled me. The air temperature in the sauna certainly felt pretty warm, and if the air wasn't being heated, why had the temperature risen? Sure, my body temperature and the air in the sauna both felt cooler than it would have been in a normal sauna, but it was still heated. So the conclusion, I guess, is that infrared saunas are cooler and, perhaps, more pleasant and less intense than ones set to higher heat. But what about recovery?

One selling point for recovery, according to one infrared sauna proponent I talked with, is that the infrared heat will somehow help your body remove a bunch of toxic substances that supposedly accumulate as a result of exercise. Infrared saunas "get the blood going quicker, so you're able to remove toxins that are built up," said the senior sales director at one company that makes and sells infrared saunas. I pressed her on what she meant about toxins building up from training. "When we produce energy, the ATP production itself does produce toxins within the body," she said, but she still couldn't explain what these toxins were. I've found no credible evidence that normal energy production creates toxins that require removal. And anyway, the body needs no special help; our liver and kidneys are quite adept at removing toxins we pick up from the environment.

I've found only one published study examining the use of infrared saunas for recovery—a small exploratory study from Finland that tested the use of a far-infrared sauna following strength exercise and running and found a tiny improvement in a specialized jump test compared to the control group.[1] Intriguing, perhaps, but it's just one particular lab measure that may or may not be relevant to most of us. Without more studies, it's hardly convincing.

"Infrared" and "far infrared" are common buzzwords in the recovery world. They're used to describe cold lasers, massage beds, and even some expensive pajamas endorsed by football star Tom Brady. There are dozens of promises floating around about the magical powers of infrared radiation—it reduces inflammation, boosts levels of growth hormones that help with muscle building and recovery, and boosts the immune system, to name a few. At best, these claims are built on tiny studies, some of them in lab animals. None of these purported benefits are definitively confirmed. Some of the claims for infrared saunas—for instance, that they are "excellent for reducing swelling, inflammation and associated pain" and that "use of the sauna expels toxins from the body"—are so egregious that the FDA ordered a manufacturer making these statements to stop.[2]

Infrared radiation is a real thing, but more often than not, the term is invoked to give an aura of space-age science to some otherwise ordinary product. You can say that a sauna is warm, or you can say that it heats your body with infrared radiation. Same thing, just different words.

———

One of the most popular ways that athletes use to enhance recovery after a hard effort is with massage. It's an old-fashioned, low-tech approach to "flushing" the muscles. The technique is almost standard practice among professional athletes. Cycling teams travel with soigneurs who rub down the athletes' legs after every ride; and nearly every NBA, NFL, MLB, and NHL training room provides massage therapists to work on achy muscles.[3] Massage is standard fare at most Olympic training centers too.

Really, who doesn't enjoy a massage? It feels really good, but despite all the love that athletes feel toward having their muscles rubbed and pressed, "There are very few evidence-based benefits for massage," says Paul Ingraham, a massage therapist, former editor at ScienceBasedMedicine.org, and publisher of PainScience.com. "There are about 100 popular reasons people give for why massage is good, and I think about 98 percent of them are nonsense." While there's some evidence that massage helps anxiety and depression, there's not much to show that it enhances performance or recovery. "Massage is a powerful placebo," he says, because it provides interesting sensations to enhance your expectation that something good is happening. "With massage, you get pleasure with expectation, and the pleasure can be quite potent. It can feel extremely good." If nothing else, there's value to lying still and feeling good for an hour—that's almost a textbook definition of relaxation. Recovery in a nutshell.

Is there evidence that it actually speeds recovery? "Massage can be effective for performance if the recovery period is short—up to ten minutes. Otherwise there's not a lot of evidence that it helps with recovery of performance," says Shona Halson, the Australian recovery expert. Massage is often explained as a way to push lactate and other waste products out of the muscle, but Halson says that "there's no evidence to say that doing massage will clear lactate—it just doesn't exist." And unless you have to race again in an hour and a half or less, your lactate will clear on its own anyway, so that's a poor reason to do it. (And since we know that lactic acid isn't the culprit behind soreness, it doesn't make so much sense to fixate on it anyway.) Sure, massage makes athletes feel better, but it's probably not related to blood flow, she said.

If massage helps with recovery, it probably does so by some mechanism other than flushing, says Timothy Butterfield, a researcher at the University of Kentucky. A couple of years ago, Butterfield's colleague Esther Dupont-Versteegden had a flash of insight into one possibility. A lot of what resistance exercise does to the muscle is related to mechanical forces, and she wondered if the same might be true of massage. Dupont-Versteegden is a health sciences professor and director of the rehabilitation services doctoral program at the University of Kentucky, where she studies muscle atrophy. Her research with Butterfield has shown that massaging a muscle can produce changes in immune cell markers, and these changes seem to vary depending on how hard and deep the massage goes. "How that changes pain, we don't know," she says.

Butterfield's studies on massage have turned up intriguing evidence that massage performed shortly after exercise may increase protein synthesis in the muscle, at least in rats. "We think it's the mechanical signal that the massage gives. We're giving the cell a signal to react differently." His results have been repeated in animal studies, but await confirmation in humans. If this happens in people too, it might mean that massage could help muscles heal from exercise-induced damage by promoting repairs via protein synthesis, but that idea remains theoretical and unproven as of yet.

But even if massage doesn't directly enhance recovery or performance, it potentially has other important benefits that are subtle, hard to measure, and almost impossible to prove. Many athletes (me among them) report that getting massage helps them to have a better relationship with their body and become more self-aware and in tune with how their muscles are feeling. "These are probably the reasons that athletes feel you couldn't pry their massage therapy away from them with a crowbar," says Ingraham, the massage therapist. "They want that experience and that body awareness, regardless of whether it has any proven effects on recovery or performance."

Athletes tend to prefer vigorous, rather than gentle, massage, perhaps because they believe that a deep massage will remove more lactic acid. But some caution is warranted. "Lots of people think that massage has to hurt to be effective," Ingraham says. But a massage that's too vigorous or deep can cause a condition called "postmassage soreness and malaise" (PMSM), which produces a set of flu-like symptoms. Some massage therapists explain this away as a sign that the massage has flushed "toxins" from the muscles, but this idea is totally bogus, Ingraham says. There's absolutely no evidence that there are toxins in the muscle that need to come out. Rather than "liberating" toxins, massage might actually produce waste products if it's so vigorous that it causes minor damage to the muscle. After Ingraham wrote about PMSM online, he started receiving letters from people who'd had deep massage and developed rhabdomyolysis, a serious condition that happens when muscle damage releases the protein myoglobin into the blood, potentially causing kidney damage. He says it's still an unproven hypothesis, but it seems possible that very deep massage can cause tiny tears or injuries to the muscles that release myoglobin and lead to "rhabdo."

What about that painful massage ball and all the lacrosse balls and foam rollers I'd tried at Denver Sports Recovery? Many athletes swear by foam rolling and other forms of self-massage, which are said to target fascia. Fascia is simply the connective tissue that wraps around your muscles and tissues like plastic wrap. The term can be confusing, because it's sometimes used to describe other connective tissue, such as the sheath around a tendon or joint capsules. But technically fascia is the sheath enveloping the skeletal muscles, says Jan Wilke, a professor of sports medicine at Goethe University in Frankfurt and perhaps the world's leading expert on fascia.

Anatomists once thought of fascia as passive tissue that just served as protection, but that view has changed in recent years, Wilke says. Although it doesn't have the contractile strength of skeletal muscle, fascia can actively contract and become stiff. It may turn out that the feeling of stiffness that we sometimes feel in the morning upon waking is from the fascia, he says, but right now that's just a working hypothesis. The study of fascia is still in its infancy, and the ideas that seem highly plausible now may turn out to be wrong or incomplete. But the latest research has turned up some interesting findings. "We know that fascia has the physiological characteristics to be a pain generator," Wilke says. When researchers injected a hypertonic salt into the fascia (a harmless way to inflict pain in the lab), people rated it as more painful than when the injection was made into the muscle.[4] This could well be the reason that it can hurt to use a foam roller or one of the countless other self-massage devices that have become popular.

The idea behind foam rollers and the like is that they loosen up the muscles and that they might address adhesions that may form between layers of fascia. "The tissue can stick together because fascia is not only a sheath around the muscles, it also has three or four layers and between the layers we have hyaluronic acid—the same fluid that's in joint capsules," Wilke says. The hyaluronic acid reduces friction to allow the layers of fascia to glide and move smoothly. If you don't move, it's possible that the fascia could become sticky and restrict movement, and if this theory is correct, it might explain why some people wake up in the morning feeling stiff, Wilke says. One idea behind pressing and massaging a muscle group with a foam roller or similar device is that it works out adhesions between layers of fascia. "This is a very very plausible theory, but it has not been proven so far," Wilke says.

And then there's the brain theory—the idea that applying compression to the muscles with a foam roller or similar device sends a signal to the brain that tells it to decrease the excitability of the muscle and relax the muscle tone. "If the tone of the muscle is lower, it might have a better range of motion," Wilke says.[5] He's "quite convinced" that there's a neural mechanism behind the benefits people report after foam rolling. There's fairly good evidence that people who do foam rolling after exercise report less muscle soreness afterward, but whether the difference is merely a change of perception or whether it's really changing something in the muscle remains unclear. A 2017 study found that foam rolling the back of one leg also increased ankle flexibility in the other one.[6] Research by David Behm at Memorial University of Newfoundland has similarly suggested that foam rolling, say, the right calf muscle decreases the pain threshold in the left calf too.[7] Results like these suggest that there must be an important neural component to what's happening, Behm says. Still, the research on the benefits remains slim.[8] "It's a good placebo. If you do it and you think you're doing something good for your body and you believe strongly in it, it will work," Wilke says. It's early days in this research, but as the science unfolds, we should learn more about the physiology behind why massage and rolling makes us feel better.

———

My dog was barking to alert me to the neighbor knocking at the door, and I was in a panic. I knew she was coming by, but I couldn't come to greet her because I was lying on my bed with my feet flailing up in the air, trying to extract myself from a pair of compression tights. These Zoot recovery tights were made of a tightly woven synthetic fabric, and I'd had to shimmy myself into them like a pair of very skinny jeans. As challenging as it had been to get them on, I was now discovering that they were even more difficult—no, nearly impossible—to take off.

I'd lifted some weights and then gone running earlier that morning, and the tights were supposed to help my muscles recover. The fact that I'd had to work a little bit to get them on didn't worry me at the time. I figured it was a sign that they'd be snug enough to really work. I'd worn the tights for nearly two hours as I lounged around the house on that Saturday. Initially they felt okay, like a pleasant squeeze around my muscles, but after a while, they started to feel constricting, particularly behind my knees. I began to slide them off, starting at the waist, and everything went fine until I got to my big fat calves. The tights were now stuck and I was doing a ridiculous dance to try and kick, pull, or peel them off.

As I heard my neighbor call out to me, I considered getting the scissors to cut the damn things off, but then I realized that to reach them, I'd have to shuffle past the glass front door in my underwear, with these stupid tights dragging along behind me, clinging to my calves. Eventually my neighbor gave up and left, and shortly after, I finally managed to extract myself from the tights. I can say with certainty that the saga did not help my recovery.

Compression tights, socks, and sleeves are everywhere these days, and they come in several styles. One variety, commonly seen at marathons and basketball games, is designed to support the arm or leg muscles during exercise and reduce muscle vibration and fatigue. These products, which include things like Copper Fit and 2XU, have become ubiquitous on track ovals, sports fields, and courts across the globe. Recovery versions, like my unremovable tights, are intended to "facilitate the return of blood back to the heart, which will reoxygenate it," said Nick Morgan, a sport scientist and founder of Sports Integrated who consults with several companies that make compression wear. "It's a flushing system, helping the body move back to baseline." Well, yeah. Normal circulation brings blood back to the heart, which pumps it to the lungs where it's oxygenated. But what are the tights doing that my body wouldn't be doing without them? To his credit, Morgan did not say that compression clothing would remove lactic acid, but he did say that it can help manage swelling. "Swelling is actually a good thing—that's what stimulates the adaptive response, but sometimes you have too much swelling and you want to reduce it," he told me. One of the biggest reasons athletes use compression garments is to help with muscle soreness, Morgan said.

As I learned, a garment's fit and degree of compression are crucial, and these can vary by brand. I've tried about a half-dozen products, including SKINS. On that company's sizing scheme, I was clearly a size small, and when the tights arrived I briefly worried about having another ordeal. (I usually wear a medium.) Instead, I found them easy to get on and off, but maybe that wasn't such a great thing. The level of compression they delivered seemed nominal. They were comfortable, but the opposite of the tights that wouldn't let go. If they were compressing, it wasn't much. As I shopped around, I found lots of other brands that had the word "compression" on the label, and most of them felt like ordinary tights. Morgan has tested and consulted for numerous brands and confirmed that the level of compression is anything but uniform across products. Your body is uniquely shaped, so it's tricky to predict how much squeeze you'll get from a given pair of tights or sleeves since fit is so individual.

What's the evidence that wearing these things will help me recover? The studies on compression clothing are mixed. Shona Halson, the Australian recovery expert, said that "overall, if you look at the majority of studies to date, there seem to be small positive effects from compression garments for both performance and recovery." A 2013 meta-analysis found that compression clothing moderately reduced delayed-onset muscle soreness, as well as aiding the recovery of muscle function after exercise.[9] A 2017 review found similar small benefits.[10] Morgan said that there's a psychological component at play here too. Some people really like the way compression garments feel. During exercise, they can reduce vibration in the muscle, and afterward they can make muscles feel that they're getting a hug.

———

In the 1990s, Laura Jacobs was the chief of physical medicine and rehabilitation at Cooper Hospital University Medical Center in New Jersey, and she regularly saw patients with terrible circulation problems. A tinkerer by nature and an engineer as well as a doctor by training, Jacobs thought she might come up with something to help these patients. "She thought, okay, what's the science? How is a healthy body moving fluids throughout the body and can we mimic that?" said her son, Gilad Jacobs. (Laura died in 2012.) When you've got blood that's pooling in the feet, the pump action of the calf muscles acts like a little heart, pushing blood back up the veins and lymphatic system and holding the pressure in a way that prevents the fluid from falling back down to the feet. She realized that it would be possible to create a pneumatic cuff that would be placed on the feet and legs to mimic this action. After some trial and error (and a lot of ideas sketched out on Dunkin' Donuts napkins), NormaTec was born. The device and company are named after Gilad Jacobs's grandmother Norma. "We're just glad her name wasn't Mildred or Phoebe, it wouldn't have the same ring to it," Jacobs told me.

The NormaTec compression system has attachments for legs, hips, and arms. The device uses compressed air to inflate balloon-like attachments around the particular body part. The idea is to increase blood flow and reduce swelling by helping the lymphatic system transport fluid from the muscle into the blood. Using a pulsating, graded pattern of compressing the muscles, the device squeezes the fluid away from the extremities and toward the heart, where the fluid can be circulated.

The NormaTec devices "allow the blood to flow easier and drive the bad stuff out," Jacobs said. What bad stuff? "That's a hotly debated topic," Jacobs said. "Lactic acid was the buzzword for twenty years, and then people said maybe it's about calcium. It's still an emerging science." In Jacobs's view, "anything we can do to work the tissue and speed up the blood flow is advantageous." Maybe, but whether pneumatic devices really speed things up remains unclear. It's one thing to use such a device on a person with poor circulation (the NormaTec system was originally made to improve blood flow in people with circulatory problems), but most athletes aren't suffering from impaired veins. While some studies on pneumatic compression devices have found improvements in blood flow and reduced swelling among users, others have found no change. It's biologically plausible that pneumatic compression helps muscles recover, but it's also likely that it's largely psychological. Athletes like it because it feels really good. "It's nice to feel a pressure sensation instead of a pain sensation," Halson says.

If I had to guess, it's this pleasant feeling that's made pneumatic compression systems like NormaTec a must-have item for athletes from nearly every sport. Olympic gold medal–winning gymnast Simone Biles used them before the Rio Olympics, and other users include Golden State Warriors player Kevin Durant, CrossFit Games champion Annie Thorisdottir, Australian cricket player David Warner, the US Olympic karate team, and the Phoenix Suns NBA team. They're hands-down one of the most coveted recovery toys among the professional athletes I've talked with. Although NormaTec is the most well-known brand, it has numerous competitors. At several thousand dollars a pop, they're too expensive for most people to buy for themselves, but many pro teams purchase units for their athletes, and they're also available at many gyms, recovery centers, and even at the finish of some triathlons and other sporting events.

———

Having tried them out, I've confirmed that blood-flushing/circulation-boosting modalities make you feel good. The compression boots I tried at Denver Sports Recovery are one of my favorite recovery toys, but I'm still puzzled by the widespread notion that athletes need special gadgets to facilitate blood flow. By nature of what they do, athletes already have good blood flow. If there's something impeding an athlete's recovery, circulation probably isn't it. In fact, we've known all along that one of the simplest and most effective ways to increase blood flow is with exercise. When you move, you raise your heart rate, and the flow of blood through your circulatory system speeds up too. If you want to promote blood flow through tired muscles and help them clear by-products of intense exercise, such as lactate, a simple, effective way to do this is with easy exercise like the gentle spinning on their trainers that Tour de France riders sometimes do after their races.[11] This approach is called "active recovery." Or, as my high school track coach used to call it, a warm-down.

6

Calming the Senses

Optimal recovery requires a certain state of mind—one that doesn't always come naturally to highly driven athletes. I once had a coach tell me, "Any fool can go train more. It takes courage to rest."[1] Trond Nystad, then–head coach of the US ski team, was referring to a common problem among athletes, who tend to be a self-motivated bunch. We get antsy and anxious when we aren't out doing our sport. It feels good to push ourselves, and we've been taught that hard work translates into success. It's tempting to conflate resting with quitting or going soft. But as Nystad explained to me, sometimes the path to performance requires doing less, not more.

I see it again and again, especially among my runner friends. We get a cold or a nagging pain in the knee, and instead of acknowledging it, we try to pretend that it's nothing. If we back off at all, it's done according to a presupposed timeline based on wishful thinking—okay, I'm sick today, but I'll be fine in two days, at which time I'll go out and train twice as hard to make up for lost time. This is how I've turned a simple cold that should have sidelined me for less than a week into a lingering illness that knocked me out for several weeks, and a minor hamstring injury that I might have overcome by taking a couple of weeks off into a performance-blunting injury that lasted an entire season. To master recovery, I had to learn to chill out.

———

I'd never met the guy, but I felt a pang of resentment toward NBA great Stephen Curry as I lowered my naked body into a small oval pod filled with salt water. I had come to the Reboot Float Spa in San Francisco's Marina District a few hours after a leg-pounding eight-mile trail run, because the star shooter for the Golden State Warriors had credited "floating" in one of the salt tanks at Reboot with helping him relax and recover during the year he became the first player in NBA history to hit four hundred three-pointers in a single season. An ESPN segment featuring Curry and his teammate Harrison Barnes waxing lyrical about the benefits of floating helped the technique become the next big thing in recovery.

Cyclists, yogis, triathletes, UFC fighters, CrossFitters, and Silicon Valley life-hackers have also gotten in on the trend. Even Homer Simpson has tried floating. In an episode of *The Simpsons*, Lisa Simpson drags her dad, Homer, to a New Age float center.[2] After an anxious start, where Lisa says to herself, "Hey, it works! Oh no—that's thinking . . . ," Lisa sees a kaleidoscope of fanciful images, while asking herself, "How am I supposed to hallucinate with all these swirling colors distracting me?" (Meanwhile, Homer is in an adjacent tank, muttering to himself, "Boring!")

Homer may not have fallen in love with floating, but Curry claimed it helped him relax and "get away from all the stresses on the court and in life." After hearing Curry and many other athletes praise the benefits of floating, I felt compelled to try it. I was sure I'd hate every moment of the hour-long session I'd signed up for at a discounted price of $60.

Back when float tanks were best known for their appearance in the 1980 film *Altered States*, about a mad scientist studying psychedelic experiences, these self-contained pools of salt water were called sensory deprivation chambers. "Sensory deprivation" sure sounds like a form of torture, so it's not surprising that the latest iteration of these devices has been rebranded as "float tanks." (Don't think of it as being locked in a cramped, dark space; imagine it as floating gently through space.) The idea is simple—you lie in a small, dark chamber filled with a shallow pool of salt water. The salinity of the water allows your body to float weightlessly, and the dark and quiet block out distractions so you can relax your body and quiet your mind. At Reboot, I'd be floating in a pod filled with about 200 gallons of water and 1,200 pounds of Epsom salts. The thought of it made me so anxious that I almost backed out of my appointment.

Floating was pioneered in the 1950s by neuroscientist John C. Lilly as a way to treat behavioral disorders.[3] Lilly spent many years studying dolphin–human communication, and he came to view floating as a way to nurture self-awareness and "personal harmony." In 1977, he published *The Deep Self: Profound Relaxation and the Isolation Tank Technique*, which explored how floating in a sensory deprivation tank (as the devices were then known) could expand a person's awareness of their internal states of being. *Altered States* is loosely based on Lilly and his experiments with the tanks (and with psychedelic drugs).

The introductory video that I watched at Reboot explained that the experience would put me in "a state akin to being in the womb," which people often find "extremely euphoric." The video told me to think of floating as "a reset button, physically and mentally," and the Reboot website claimed that floating "lowers stress hormones, replenishes neurotransmitters, and releases endorphins, which induces ultra-deep relaxation and provides a zen-like afterglow that can last for days."

As enticing as such euphoric relaxation sounded, I felt anxious. The thought of disrobing and closing myself inside a dark pod only slightly larger than a coffin made me claustrophobic. Stillness is not my natural state, and my monkey brain has a habit of hijacking my meditative mind. I braced myself to spend an hour in the tank, hoping to just get through it so that I could describe it and then never go back.

My private float room was equipped with a tile shower and a float pod, which looked like a giant white clamshell. The lid lifted and closed on pneumatic risers, and inside was a light with three settings—psychedelic purple, a slow fade from purple to blue to green and orange, and off. My hour would begin with a few minutes of soft instrumental music, which would gradually fall out and then return to indicate when I had five minutes left. (When time was up, the pod's high-powered filter would kick in, to cleanse the water.)

I opened the top and slipped inside, leaving the lid ajar to prevent myself from totally freaking out. The water was body temperature—neither warm nor cold. I felt wondrously buoyant in the water, which was so salty that the spa provided a spray bottle full of fresh water and a wash cloth to wipe my eyes in case I got salt water in them. Though the water was just a few inches deep, I easily floated without touching the bottom. I had to experiment a little to find a position that felt natural. Eventually I settled on letting my arms drift above my head as if I was raising my hands.

The guy who'd showed me around had urged me to try at least a few minutes with the light off, and as he said that I thought, *Yeah, right—there's no way I'm doing that.* But once I was in the pod, my attitude relaxed along with my body. After maybe 10 minutes of the different colors fading in and out, I hit the switch and the light went off. I'd intended to hit the button twice, to turn it to a constant purple, but instead I banged it once, to off. I reached for the switch again, and when I didn't immediately find it, I thought, okay, why not try a minute without the light?

By this time, the music had stopped, but I didn't miss the stimulus. It was quiet and dark, and all I wanted to do was give myself over to this gentle euphoria. It was like I was floating in a stream of consciousness, but instead of pulsing with nervous thoughts or energetic ideas, my mind was simply drifting into a pleasant nothingness. There were no demands on my attention, none of my senses were being stimulated, and the tension in my body seemed to melt away. I'd done a leg-pounding trail run just a few hours earlier, but now my muscles and joints felt totally loose and relaxed. For a period that seemed timeless, I could just be. It felt like an extended version of that moment when you're falling into a deep slumber.

Beforehand, I wasn't sure I'd last a whole hour in the pod. But when the gentle music returned to indicate that I had only five minutes left, I felt a moment of panic. It couldn't have been an hour already, I thought. *I'm not ready to get out!*

The prefloat video had made a lot of promises that, at the time, seemed overblown. But by the time I got out, I was sold. I didn't have any hallucinations or memorably deep thoughts, and I didn't believe for a minute that my body was "detoxifying." I was skeptical that floating had flushed out lactic acid or reduced inflammation in my muscles or lived up to any of the long list of claims. But I was certain that I'd just experienced sixty of the most relaxing minutes of my life. A whiteboard in the hall-way had the words "Floating makes me feel . . ." written in big letters across the top with answers scribbled in different colors and handwriting below: *supreme, floaty, amazing, spacey, awe-some yo, gnar, relaxed, tranquilo, whoa.* Floating made me feel serene. That night, I would easily fall into a deep and satisfy-ing slumber, despite the long day of travel and the challenges of sleeping in an unfamiliar place.

As I left, the attendant asked when I'd be coming back. "Tomorrow?" I replied. Screw the board meeting I'd come to San Francisco to attend. I wanted more time in the pod. So did a lot of other people, apparently. He told me that they were booked several weeks out.

———

The early scientific models of training, recovery, and adapta-tion all but overlooked the role of psychological stress in the process. But research beginning in the 1970s suggested that the body's physiological response to exercise depends at least in part on one's emotional state and perceived ability to cope with the challenge. "So much of the focus has been on the physi-cal aspects of recovery, and the psychological aspects have been completely neglected," says Jonathan Peake, a sports scientist at Queensland University of Technology in Brisbane, Australia. General feelings of health and wellness play an important role in how people recover and adapt to training. "This could well be the exciting frontier of recovery research," Peake says.

As far as the body is concerned, stress is stress—it doesn't matter if it comes from a session of intervals or from the emo-tional strain of a romantic breakup, says John Kiely, an Irish sports scientist and performance coach who's worked with world-class athletes in numerous sports, including rugby and track and field. "If you really want to optimize recovery, you need to manage stress," Kiely says.

Psychological stress doesn't just impair recovery, it may also blunt the body's ability to adapt to training. A 2012 study asked volunteers to rate their mental stress and their resources for coping with the stress before they began a supervised cycling program designed to improve their fitness.[4] The study only lasted two weeks, but the participants hadn't previously been exercising, so they were expected to show some pretty immediate gains. (It's easier to improve when you're starting from zero than when you're already highly trained.) Results showed that people who reported low stress levels posted a marked boost in aerobic fitness and maximum power, but these improvements were small or absent in the people who reported high stress levels. The study hints that stress may be a contributing factor to variation in physiological responses to exercise.

Mental stress also seems to increase the odds of an injury. A 2015 study of Division I college football players found that during times of high academic stress the risk of an injury was almost twice as great as during times when school-related stress was low.[5] These differences were most pronounced in athletes who regularly played in the games, leading the authors to speculate that high academic stress "may affect athletes that play to an even greater extent than high physical stress."

Athletes often think of recovery only in terms of planned physical exercise, and they overlook the emotional strains on their bodies. "Your training isn't just, go out and do your run. It's what you do after your run too," says Kiely, a former elite kickboxer. A classic example, he says, is the athlete whose rest day is spent scurrying about doing emotionally demanding activities like household chores, tax accounting, or frustrating errands. "It ends up being quite a stressful day. Then you have to ask, was that really a rest day or just another day where you overlaid some other stresses?"

When it comes to stress, what matters is the potential stressor's emotional content—how stressful it *feels*. And how stressful it feels depends largely on expectations, Kiely says. "If I expect to be able to handle this stress, then I can. If I expect that I can't handle it, then this expectation will amplify it." What's stressful for one athlete might not feel stressful to another. For instance, grocery shopping might feel enormously trying to an athlete who is struggling to eat well on a tight budget, whereas it might provide a relaxing pleasure to a foodie who enjoys cooking. A person's personal circumstances, their genetics, and even their upbringing can influence how they process and experience stress. These inclinations aren't set in stone, Kiely says. "We can't change an athlete's genes or fetal environment or nurturing, but we can change how they experience stress by helping them to reframe things and giving them a toolbox of stress-handling capabilities."

This stress toolbox might include listening to some favorite music, reading a poem, meditating, or taking a walk in the woods. "The basic rule is, it has to be something you enjoy," Kiely says. "I go to the athlete and say, 'Let's find a tool that fits you.'" Mental relaxation is rarely prescribed with the same degree of precision that training is, but it should be, Kiely says. Sometimes the prescription is as simple as, go outside and walk your dog. Seriously, he says. "Pet owners have less stress reactivity."

I have long believed that the most important part of my day is the walk up the hill and through the woods that I take every morning with my husband and my dog. The ritual connects me to the two of them, to nature and my place, and it's a chance to move my body, clear my mind, and reflect on the day. It's physical activity that feels relaxing, rather than taxing, and it allows me to gauge how my body is feeling.

I once shared this observation with Kelly Starrett, the renowned physical therapist and CrossFit coach, and he nodded his head in recognition. "You know, one of the most important performance behaviors [my wife] Juliet and I have changed is that we walk our kids to school every day. We start our day with about a mile and a quarter walk to school, and already that's a radical change. Our nonexercise activity goes up, and by the time I roll in to work I'm already warm and feeling good. And that's all from a low-tech thing called walking."

———

In the early 2000s, I was zooming around North America competing with the Rossignol ski team. Traveling with a load of ski equipment in the winter regularly involved lost bags, weather-delayed flights, and other petty injustices. After one particularly rough journey where my race skis went MIA, I was tired, hangry, and on the verge of a meltdown. Our team director took me aside for some words of advice. Look, he said. These hassles are just part of the job. You want to be a ski racer? Then learn to deal with it. You can waste all your energy getting pissed at the airline, or you can accept that this stuff is going to happen and save your emotional energy for the race. He was right. I couldn't change what was happening, but I could change the way I responded. That shift in attitude instantly reduced my stress levels.

Life presents many situations where stress isn't avoidable. In these instances, it's pointless to try to eliminate the stressor, says Göran Kenttä, a sports psychologist at the Swedish School of Sport and Health Sciences. Instead, the solution is to change how you relate to the stress. Kenttä worked with the Swedish swim team at the 2016 Rio Olympics, where two of the women competing had young children. Some of the parenting-related stresses these moms faced on a day-to-day basis weren't going to go away, so the trick was to find a way to coexist with them.

"Learning to respond to that kind of stress is about changing your relationship to the stressor," Kenttä says. It may sound counterintuitive, but he says that rather than turning away from a source of stress you can't eliminate, it's more effective to turn your attention *toward* it. "It really boils down to one thing—if there are thoughts that you don't dare to have, you're stuck with them, that's the bottom line," he says. "It's accepting [the stressor], and accepting it just as it is." With that awareness and a mindful state, you can decide what you want to do with the situation. In the case of my relentless travel hassles, I stopped dreading them and instead expected and, to the extent I could, prepared for them. I couldn't help that my preferred food wasn't easily available in airports or at rest stops, but I could travel with a cache of healthy snacks. I couldn't do anything about it if I arrived at a race and my skis didn't, but I could accept that this was out of my hands and make a plan B (to race on a pair of team skis).

Travel is a common source of stress among athletes. In January 2017, ultramarathoner Mike Wardian competed in the World Marathon Challenge—a ridiculous event that has runners complete seven marathons on seven continents in seven days.[6] Wardian shattered the event record by averaging 2:45:57 over the seven courses.[7] The Arlington, Virginia–based runner had plenty of prior experience with long-distance events. He's won numerous 100-mile runs, has been national champion at 100 km and 50 miles, and holds the world record for fastest marathon run while dressed as Elvis Presley. What made the World Marathon Challenge particularly, well, challenging, he says, was the relentless travel between events and short rest periods between races. The week followed a pattern: arrive at the new country, run, then get on a plane and go to the next continent. "My sleeping was terrible," he says. In many cases, his best opportunity to sleep was on a plane—not the most conducive environment for sleeping.

For help, he turned to meditation, a technique he'd recently adopted after hearing a plug about a meditation app called Headspace on a favorite podcast. "Meditation is one of those touchy-feely things, but this didn't seem like this app was like that. It was very clever. They suckered me with a video and this guy's voice and it was free, so why not?" He listened to the app's ten-minute guided meditations, which walked him through the process. The app didn't always put him to sleep on those plane rides between continents during the World Marathon Challenge. But, Wardian says, "it made me calmer, and it felt like maybe my body was better able to heal itself." He likes the way meditation makes him feel, so he's stayed with it. He does it every few days, usually before bed, but sometimes in the morning before getting out of bed for a run.

Headspace has received lots of attention. Its British founder, Andy Puddicombe, a Tibetan monk turned TED Talk guru, has been profiled in the *New Yorker* and elsewhere. I've talked to several coaches who have encouraged their players to try the app. It's apparently popular among some NFL players, and a recent ad campaign (yes, meditation is now an advertised product) featured a power lifter who says, "I meditate to crush it."[8]

Puddicombe's breezy meditations weren't for me, but I wasn't ready to dismiss meditation yet. I hoped it might help me reach the kind of meditative bliss I've achieved in the float tank. So when I learned about a headband that teaches people to meditate by providing biofeedback based on their brain's electrical activity, I was intrigued.

The headband is called Muse, and it uses electrodes and electroencephalography (EEG) to measure the electrical activity in your brain. During meditation, experienced meditators tend to have a shift in their brain activity in favor of moderate-frequency theta waves, and the device promises to translate what's happening in your brain when you meditate into "guiding sounds" that can help you stay on task. "This headband is amazing," Peake, the scientist studying exercise recovery at Queensland University, told me. "It provides biofeedback to train people to meditate. We think it has great potential." Muse purports to help you notice the subtle shifts in your mental state so you can more easily learn to tamp down excitable thoughts through meditation.

You wear the headband across your forehead and tuck the flares at the end behind your ears. Electrodes on the front of the headband and the ear flaps are supposed to measure brain waves—electrical currents flowing in your brain. (The device pairs with a smartphone, which provides the audio feedback.) Each session begins with a "baseline" reading, and the first time I tested it, it took about four tries to get the headband to work. After a bunch of fiddling with the placement, I finally got my baseline, and then I moved on to a ten-minute session. Before I started, I had a chance to pick a soundscape—everything from ambient music to desert, rain forest, or beach. I selected rain forest and was off.

Muse's trick is that it provides sounds that respond to what the device is measuring in your brain waves. If your brain seems calm, so are the waves or the gentle rain. But if the EEG indicates that your brain is in a more active state, the sounds become more agitated or stormy. If you become really relaxed, you hear birds fluttering quietly in the background. At the end of the session, you're presented with a rundown on how you did—how many minutes your brain was "active," "neutral," or "calm." You also get points for calm time and a bird count.

I don't like what it says about me that I found this quantification of my meditation session appealing, but I did. I also found it deeply disconcerting that meditation could be "gamified." "One of the first things we worried about was addiction. Are we going to turn meditation into Candy Crush?" Jay Vidyarthi, former head of user experience at Muse, told me. "That concern dissipated pretty quickly. Meditation is actually really hard, so it was like asking, is FitBit going to turn running into Candy Crush?"

He was well aware that by making meditation into a kind of game, Muse introduced an element of judgment. But "what we sacrifice with that, we gain in motivation," Vidyarthi said. "Muse is an act of compassion bringing meditation to an audience that might not otherwise access it." It's not Muse users' fault that they're achievement-driven, he said. "We don't do competitive leaderboard, and you're only competing with yourself. It's up to you whether you do that with integrity." The goal for beginning meditation is not to be in a perfect, harmonized state, Vidyarthi said. "It's to build your ability to pay attention."

Muse is designed and intended as a training tool. The feedback it gives helps you know if you're achieving the desired state of focus, and it lets you know when your mind is wandering (in case you can't notice that on your own). After a few sessions, I found myself getting the hang of it. I wanted to hear birds, and if I didn't hear them, I would focus on breathing more deeply and on keeping my attention on my breathing. But I also noticed that sometimes I felt extremely relaxed and unencumbered by stray thoughts, even when the soundscape was a bit agitated and bird-free. One morning before work, I felt particularly relaxed and decided to do a short session, imagining that I could get a high score and rack up a bunch of birds. (I realize that I was missing the point.) This is how I discovered the importance of the baseline measurement. It turned out, when I started from calm, it was very hard to improve, and the baseline was what my meditation session was being measured against. Because I'm a scientist at heart, I did some experimenting and found that the more active and agitated I had been during the baseline test, the more positive feedback I'd received during my meditation session. Simply opening and closing my eyes repeatedly during the baseline could throw my score so that I could spend my session blissing out to a chorus of birds.

I asked Vidyarthi about this, and he acknowledged that the baseline can skew things, and added that they're working to address this by collecting more longitudinal baseline numbers. But the issue raised another question too—how accurately does the little plastic headband really measure EEG and how meaningful are the data?

"Can you pick up on an intensity of concentration with EEG? I think you can, but the Muse is going to be a pretty low-resolution method of doing that," says Norman Farb, a psychologist at University of Toronto Mississauga who has done some research with the Muse headband. A product like Muse has fewer electrodes than the devices used in the lab, and so the data produced will be more crude. Whether EEG readings translate to mind states isn't completely straightforward. Not every experienced meditator shows the same brain wave patterns, so there's plenty of squishy interpretation. What perhaps matters even more is whether the meditation sessions are having the intended results. Much of the research that's been done on meditation has focused on people with extensive training, and it's not clear yet whether devices like Muse and apps like Headspace can deliver the same kind of results as the more traditional eight-week guided course. "You have this basket of positive expectations. Are you really selling something more than a basket?" The jury is still out, Farb says.

Sports psychologist John Sullivan works with players from the NFL, NBA, WNBA, and MLS as well as some Olympic athletes, and he makes the case that athletes are trained to seek feedback. Although ideally meditation should be done without judgment, he said, "from a sport standpoint, that doesn't hold. Athletes are constantly being judged, and they want to see their data." Sullivan uses Muse with some of his athletes, and what makes it appealing is that it offers actionable information.

After playing around with Muse, I've settled on a meditation practice that uses a 15-minute yoga audio recording that begins with instructions to relax into the shavasana pose, then a couple minutes of a soothing voice reading a poem, followed by instrumental music. I've found that listening to the same calming recording each time serves as a cue that it's time to relax and regroup. I don't need to focus on instructions, I can put my attention on my breath and just observe my thoughts float by, without judgment. The point is to become an observer of my thoughts, rather than the hand that churns them, and I find this is easiest to do when I'm alone in my head.

Athletes and members of the military serving as study subjects in the air force's Signature Tracking for Optimized Nutrition and Training (STRONG) program have tested a lot of different ways of enhancing recovery, and they've found one that seems to work for everyone. "The float tank has been home run recovery for us," says Joshua Hagen, then STRONG's team lead at the Air Force Research Laboratory just outside of Dayton, Ohio (he is now the director of the Human Performance Innovation Center at the Rockefeller Neurosciences Institute at West Virginia University). "I can put any athlete or operator in there and it's going to make them better." A study Hagen's group conducted with researchers at the University of Cincinnati found that floating can rebalance the sympathetic and parasympathetic nervous systems, which tend to get thrown off kilter when someone is stressed or fatigued. (The parasympathetic system promotes relaxation and controls subconscious activities like breathing and digestion, while the sympathetic system controls your stress reactions and fight-or-flight response.) He says floating works whether the person is overstimulated or exhausted. "It's the only modality we've seen that can fix you either way." He says that floating has been used in the military for a long time, and "the guys love it."

I fell in love with it too. Back home in Colorado, I found two nearby places where I can float, and after a half-dozen or so sessions, I've come to think of it as forced meditation for people who aren't naturally inclined to meditate. That may explain its popularity with military guys and macho jocks. "You're alone with your thoughts. You have to shut your brain down, and we never shut our brains down," Hagen says. "You have to relax, and your body is inevitably going to undergo some good recovery." The STRONG team has a collaboration with the University of Cincinnati to study the performance and recovery benefits of floating. The full results aren't in yet, but one measure—of the stress hormone cortisol—looks promising. "We're seeing a 25 percent reduction in blood cortisol before and after a float," Hagen says. A study published in 2016 examined the effects of floating on sixty elite athletes from nine sports and found that floating reduced perceived muscle soreness and improved participants' moods—making them feel more relaxed, calm, and happy, and less worn out and tense.[9] Another study of people recovering from stress-related illnesses showed that floating improved their sleep.[10]

Floating isn't just a method of stress reduction, it also can help athletes hone their mental focus. Sessions in a float tank were an important part of cyclist Evelyn Stevens's preparation for her run on the world hour record—a measure of how far a cyclist can ride on a velodrome track in one hour. A world champion time trialist and two-time Olympian who quit a demanding job in the Wall Street financial industry in 2009 to become a professional cyclist, Stevens understood from the beginning that mental focus would be key to breaking the record. Once she was under way, she would need to stay singlemindedly on task. "It was unlike any kind of pressure I've ever experienced on a bicycle," she says.

"My husband's a very big Steph Curry fan, and he read about the floating and realized that this float spa was right next to our apartment in San Francisco, so he booked us for it," Stevens says. Floating helped her deal with stress and anxiety leading up to the attempt so she could fully recover from her workouts and arrive on the starting line mentally rested and ready to perform. "Sometimes I might visualize for part of it, and sometimes I'd just zone out. All of my anxiety would just come flying through, which was helpful, because then I didn't have it while lying in bed at night." Hour-long floats felt to her like perfect preparation for the record. "It's 60 minutes of sensory deprivation, which to me is kind of like the record—60 minutes of deprivation. It's practice—where does your mind go in those 60 minutes?" In 2016 at the velodrome at the US Olympic Training Center in Colorado Springs full of fans, Stevens set a new hour record.[11]

———

True recovery requires nurturing a recovery mind-set—one that fully honors the body's need to recuperate and senses when it's time to chill. This attitude can be hard to cultivate in a culture that's constantly bombarding us with messages to "go hard or go home." We're primed to push through the pain and do another mile or one more set. We celebrate "streakers" who run every day for years. But if your body isn't recovering from and adapting to those runs, then you're just logging junk miles that are wearing you down instead of building you up. I don't mind celebrating those impressive people who keep their exercise streaks alive, but maybe we should lionize athletes who've mastered the art of resting too.

7

The Rest Cure

Tom Brady's pajamas arrived on my doorstep in two space-age pouches that looked like they might contain Starfleet uniforms. Unzipping the silver pods, which had the slogan "Rest. Win. Repeat." debossed on the lids, I found what looked like ordinary sportswear, matching pieces in black synthetic fabric. The shirt, long sleeves with a scoop neckline and five buttons down the front, felt soft and stretchy. The bottoms, with an elastic drawstring waist and elastic cuffs at the ankles, were made of the same thin, pliable material and had a small blue crescent moon printed on the butt.

Retailing at $199.98 for a complete set, TB12 sleepwear isn't cheap, but these weren't ordinary pajamas—they were scientifically developed recovery technology, according to Under Armour, its maker, and to New England Patriots quarterback Tom Brady (number 12), its eponymous endorser. The inside of the material was printed with a hexagon pattern containing the bioceramic powder purported to give these jammies their magic. The bioceramic material supposedly absorbs body heat and reemits it as far infrared radiation. "The far infrared, when it's against your skin, ends up reducing inflammation," Brady says in a video on the Under Armour website. "Without the sleepwear, I don't really feel like I would have been able to achieve the things that I have done and hope to continue to do."

It seemed like a bunch of overwrought marketing, but if these pajamas were really the key to Brady's athletic longevity and success, I was willing to give them a try. I'm just a few years older than Brady and entering an age in which promises of youth hold a certain allure. I'd just gone for a long, hard cross-country ski and was feeling pretty beat, so I figured it was a good night for the first test.

My initial impression was that they were comfortable, at least for lounging around before bed. The garments were pretty much indistinguishable from any other Under Armour performance clothes. They looked and felt like regular wicking tights and shirts. The light fabric was soft against my skin, and even though the pajamas covered my body from neck to ankle, they didn't feel constricting or overly warm. I'd suspected that their trick was heat, which usually feels good on sore muscles. Instead, they felt strangely temperature-neutral—not particularly warm or cold. I crawled under the covers and fell asleep as usual. I didn't notice anything different about my sleep or how I felt the next day upon waking. My run the next day felt as it usually would. I didn't feel any more or less recovered than usual. I tried them on multiple nights. If the pajamas had enhanced my recovery, the benefit was too minuscule for me to detect.

"The goal of the sleepwear is to unlock better sleep," Glen Silbert, who was then Under Armour senior vice president of global product, told me. Silbert didn't explain what it was that was keeping sleep locked up, and when I asked if the garments worked by reflecting heat, Silbert said no. "It's not taking heat and returning heat. It's basically taking heat and returning it in far infrared, which is not heat, it's an energy." I was surprised to hear that, because according to the research paper that Under Armour pointed to as proof of the scientific principles behind their far infrared technology, "FIR transfers energy purely in the form of heat which can be perceived by the thermoreceptors in human skin as radiant heat." In other words, the sleepwear absorbs body heat and reflects it back as, wait for it . . . radiant heat. Just like the infrared sauna that radiates a special heat called "infrared." To be fair, the pajamas are supposedly emitting the heat at a different, lower-energy wavelength, for whatever that's worth. (Whether far infrared radiation is heat or not is a matter of semantics, but for all practical purposes the two are synonymous. We can't see far infrared radiation, we can only sense it as heat.) In the promotional video, Brady claims the FIR from the fabric reduces inflammation, but it's not clear how that works and there's scant evidence for this claim. Anyway, decreasing inflammation might actually be a bad thing if you're hoping for a training response.

Silbert told me that lab tests of the sleepwear had shown that it "helps you go to sleep faster and sleep more restfully—meaning you wake up fewer times in the night and ultimately sleep longer." I didn't notice any of these effects in my trials, but Silbert assured me that "there's real science behind it. Tom Brady was our proof of concept." The evidence, he said, lies in Brady's performance. "At 39 years old, you could argue he plays like a 23-year-old. He is the best quarterback in the game, and I don't think he's played better than this year," Silbert said of the 2016–2017 season that ended with Brady's team winning the Super Bowl.

I could only shake my head. Tom Brady probably is one of the

NFL's best quarterbacks, but he was a successful football player before he signed his name to these pajamas, and to attribute his success to the bioceramic powder in his PJs seems far-fetched. The real trick here isn't pixie dust, it's something far more powerful—sleep. Brady is known for hitting the hay early, and his 8:30 p.m. bedtime and good sleep habits are a more likely contributor to his performance and longevity than any special clothing (or the luxury mattresses he also promotes, for that matter). Still, there's a reason Brady and Under Armour are commodifying sleep—it works. Insofar as there exists any magical secret for recovery, sleep is it.

The benefits of sleep cannot be overstated. It's hands-down the most powerful recovery tool known to science. Nothing else comes close to sleep's recovery-enhancing powers. You could add together every other recovery aid ever discovered, and they wouldn't stack up. Going to sleep is like taking your body to the repair shop. While you doze, your body's recovery processes ramp up to fix the damage you did during the day and get you ready to perform again.

Throughout the night, we cycle through four stages of sleep. Stage one is when we transition from wake to sleep, is considered the lightest stage of sleep, and usually takes up about 5 percent of our total sleep time. Next comes stage two, where we spend about 50 percent of the night and which is harder to wake from than stage one and may be important for some memory processing. Stage three is the deepest stage, which is also associated with memory processing and is when the body releases substances like testosterone and growth hormones to push tissue repair into high gear. Skimping on sleep can blunt the release of hormones involved in muscle building and rejuvenation. In one small study, healthy young men who slept only five hours per night for a week experienced a 10 to 15 percent dip in their testosterone levels.[1] Stage three is also where factual learning is encoded in the brain and superfluous memories are pruned. The final sleep stage is REM, or rapid eye movement, sleep in which most dreaming happens. Procedural memories seem to get enhanced during REM sleep, so for athletes in a sport that involves motor or cognitive skills, this phase of sleep is especially crucial. REM episodes get progressively longer as the night goes on, so when you skimp on sleep, you're depriving yourself of REM sleep.

Tom Brady makes the case for sleep in *The Tortoise & The Hare & Tom Brady*, an illustrated tongue-in-cheek promotional children's book produced by Under Armour and written by the comedy group Funny or Die.[2] In the story, Mr. Tortoise sends Brady a text message ("tom brady u suck at football") challenging him to a game. As he prepares for the challenge, Brady shares his tips for good sleep, which read like a page out of an expert's guide to sleep, with some pitches for his TB12 products thrown in, of course. In one scene, Brady tells a framed photo of Boston, "This recovery sleepwear will help my body recover faster as I sleep."

Before snuggling under the covers, Brady turns down his thermostat. "There. 65 degrees Fahrenheit. My ideal temperature for restful sleep," Brady whispers to his four Big Game victory rings. "The rings winked at him, as they do every night." Winking aside, Brady's right about the temperature—65°F is cool but not cold, an ideal temperature for sleeping.

Brady also shuts down his devices 30 minutes before bedtime. With his phone safely tucked in a special box to prevent it from distracting him, Brady dons his spectacles (just for show, of course—his vision is perfect) and settles in with his favorite book. This is totally solid advice. A small study by researchers at Brigham and Women's Hospital compared the sleep patterns and circadian rhythms of volunteers who read on an electronic device before bed to those of volunteers who read print books instead.[3] It found that those staring at a screen had suppressed melatonin levels, a delayed circadian clock, and reduced alertness the next morning compared to those who'd read print books. Research suggests it's the blue light emitted by computers, tablets, and phones that disrupts our body clocks, so apps like F.lux, which progressively reduce blue light as the night progresses, can help cut the risk for those who can't live without pre-bed screen time.[4]

As he's about to doze off, Brady gets a visit from Mr. Owl, who says, "Tom Brady, don't forget your ear plugs. A quiet environment is crucial to your good sleep." More solid advice—in addition to a consistent bedtime and wakeup time, a good sleep environment free of incoming noise, light, and other distractions can help optimize sleep.

While Brady is getting a head start on rest, Mr. Tortoise is down at the bar, slamming back shots of carrot juice. No surprise, Brady wins the game by 200 points, but he's such a good sport that he shares some TB12 brand electrolytes with Mr. Tortoise and Mr. Hare after the game.

——

Sleep is when recovery and adaptation happens, and prioritizing it can help athletes flourish. The optimal amount of sleep varies by person. "I tell athletes, if you didn't set an alarm, how long would you sleep?" says Erik Korem, former director of sport science for the Houston Texans NFL team. "Here's what I want you to do—go to bed at a normal time in a dark, cold place with no alarm. When you wake up, use that as your starting point." He tells players, "You're striving to find the sweet spot. As a general rule, that's seven to nine hours of sleep per night." Korem encourages his athletes to pay attention to when they're sleeping, though it's up to them to decide whether they want to formally track it. "The biggest thing is just awareness," he says. Once athletes start paying attention, "They realize, oh yeah, I only got six hours of sleep last night and this is how tired I feel, maybe tonight I'll go to bed earlier."

It's become common in some circles for people to brag about how little sleep they're getting, as if sleep deprivation is some sort of status symbol. I recently met a renowned neuroscientist who boastfully blamed four hours of sleep for his difficulty carrying on a conversation. (He was so busy with all his startups!) But neglecting his body's basic needs was really just a sign of self-delusion. Yes, a few lucky "short sleepers" are genetically inclined to thrive on only about four hours of shut-eye per night, but this condition is extremely rare.[5] "Most of the time when someone says they only need five or six hours of sleep, what that means is that their ability to tolerate sleep deprivation is better than most," says Meeta Singh, a sleep specialist at the Henry Ford sleep laboratory. "They're actually walking around with sleep debt, and have forgotten what it feels like to be awake and alert."

Research shows that people who are chronically sleep-deprived lose their normal perception of sleepiness and become poor judges of their neurocognitive performance, says Sigrid Veasey, a physician and sleep researcher at the University of Pennsylvania. When people are limited to five hours of sleep per night for a week, they report feeling very sleepy after the first night, but chronic sleep loss blunts the sleep drive, so after a night or two, the person doesn't feel as sleepy as they should. "You think you're functioning okay, but you're not," she says. You might be able to do rote memory tasks on five hours of sleep, but anything that requires trouble-shooting or complex thoughts will become increasingly difficult.

Skimping on sleep is like showing up to the game drunk. "If you only get six hours of sleep, you double or triple your normal reaction times," Singh says. She points to a study in which researchers subjected volunteers to either doses of alcohol or sleep deprivation and then gave them a battery of tests on attention, reaction time, and sleepiness.[6] The results showed that people who'd only spent six hours in bed were as impaired as those who'd consumed two to three beers, while four hours of rest was equivalent to about five or six beers. Staying up all night was like throwing back ten to eleven beers and then trying to function normally.

If you're forced to pick between some extra shut-eye or an extra workout, it's wiser to pick the sleep, Singh says. Sacrificing an hour of sleep to make a morning workout is totally self-defeating.

Without proper sleep, the body becomes fragile. Studies at the Henry Ford Sleep Lab have shown that skimping on slumber makes people more sensitive to pain, and a study of high school athletes found that sleeping fewer than eight hours increased the risk of injury. Sleep deprivation can also suppress the immune system and make you more susceptible to viruses. In one study, researchers measured the sleep patterns of 164 people over the course of a week, and then brought them into the lab where they were quarantined and given nasal drops containing a cold virus.[7] The people who'd slept an average of six or fewer hours per night during the previous week were far more likely to develop a cold than those who'd slept seven or more hours. Forty-five percent of the people who'd slept five or fewer hours got sick, while less than twenty percent of the volunteers who'd averaged seven or more hours of sleep came down with a cold.

Whole books have been written about how much practice athletes need to succeed at the highest levels, but if it's not paired with sufficient sleep, even 10 million hours of practice won't help. Sleep reinforces motor skills learning, and sleep in the initial 24 hours after training seems especially crucial—a finding that led researchers Matthew P. Walker and Robert Stickgold to propose that "It's practice, with sleep, that makes perfect."[8] After an afternoon working on your tennis serve, golf swing, or swim stroke, you need a full night's rest to make the lessons stick.

Measuring exactly how much quality sleep you're getting is difficult outside of a lab setting. Although lots of fitness trackers and phone apps promise to measure sleep duration and quality, these gadgets aren't very reliable. "They claim to estimate deep sleep and light sleep, but there's no way you can do that unless you're measuring the electrical activity of the brain," says Amy Bender, a sleep scientist at the Centre for Sleep & Human Performance in Calgary. One study showed that these trackers, which generally measure your movements to estimate how much and how deeply you're sleeping, were off by as much as one hour. Rather than relying on an inaccurate measure, Bender prefers to simply ask athletes how satisfied they are with their sleep quality. When athletes feel unsatisfied with their sleep, that's usually a sign that something is off. She also likes to have athletes track when they go to bed and when they get up to see if they're spending enough time in bed to get the sleep they need.

Some apps and sleep coaches advocate a protocol where

bedtimes and wake times are calculated based on a 90-minute sleep interval to ensure that you wake up in REM sleep, which is said to make you more alert during the day. "But the 90-minute cycle is not based on science or evidence," Bender says. Sleep cycles vary a great deal—from person to person, but also for the same individual from night to night. Whatever your usual pattern, you'll spend more time in slow wave and deep sleep early in the night following a really tough training session, and that will delay your REM cycle, Bender says. "It's extremely difficult to pinpoint when you should wake up, and it's not really that important to make sure you wake up during REM sleep. What's more crucial is sleep quantity and sleep quality."

Although his main job these days is coaching collegiate and professional runners, Steve Magness remains a scientist at heart. During one recent season, the University of Houston head cross-country coach gave his runners an app that allowed them to record things like how many hours they'd slept, how well they'd performed each day, and how they felt. At the end of the season, he crunched the numbers and found that sleep was the one thing that could predict how they'd perform on a given day. "It was the only thing that tracked. There was a direct correlation with sleep and how stressed they felt," he says. "It really hit home for a lot of them." In one study, extra sleep translated to improved performance in a muscle contraction test, and the exercise felt easier to those who'd had more rest.[9]

Paying attention to sleep is healthy, but fixating too much on numbers can backfire. Endurance coach Kristen Dieffenbach recalls a masters triathlete she coached who wanted to do everything by the numbers. Watching the athlete's performance plummet, Dieffenbach advised her to back off the training and get some additional rest. Instead, the athlete, a medical doctor, pushed back. "But my sleep monitor says I got enough sleep!" As Dieffenbach puts it, "There's a point at which how you feel matters a lot more than a number on some device."

Fitness trackers can mislead in the other direction too. Researchers in Chicago recently published a case study of a 27-year-old woman who came to them worried that she was waking up feeling unrefreshed, and noting that her FitBit data indicated that she was sleeping poorly.[10] She came to the sleep lab, where researchers used scientific-grade instruments to measure her brain waves, heart rate, and other physiological factors while she slept. It turned out she was sleeping fine, but the woman was reluctant to believe it because her tracker was telling her something different. Her fixation on her sleep data had made her unduly anxious about how she was sleeping, a condition that's become so common that researchers have given it a name—orthosomnia.

As anyone who's found themselves awake late at night knows, sleep has a powerful psychological component that can create feedback loops. For instance, insomniacs tend to overestimate the time they've been awake at night. "They underestimate their sleep," says Meeta Singh, the sleep scientist at the Henry Ford Sleep Disorders and Research Center, "and when they come to a sleep lab they think they've slept 20 percent of the night when they've actually slept 60 or 70 percent." The person wakes up and then feels anxiety that feeds on itself. "The feeling of awfulness when you're looking at the clock at 2:00 a.m.—that very act recruits more brain cells and wakes them up and makes it more difficult to go back to sleep." Singh tells the athletes she works with to avoid looking at the clock in the middle of the night, and if necessary, to turn the clock so it's not visible while they are lying in bed.

For athletes who have trouble falling asleep, Bender recommends the "cognitive shuffle" technique. Begin by thinking of
a word: say, "bedtime." Then start with the first letter, in this
case B, and imagine all of the words you can think of that start
with that letter. When you run out of words for B, move on to the
next letter, E, D, T, and so on. This technique aims to distract
your mind from worrying about falling asleep.

Fixating on one night's poor sleep is unhelpful. Bender tells
athletes to think of their sleep in terms of a weekly budget so
that they don't stress out over one bad night. "If you need to,
take a longer nap one day or sleep in on another. Focus on your
weekly need, rather than being concerned about eight hours
every single night."

What someone believes about their sleep may become a self-
fulfilling prophecy. In one study of more than one hundred
volunteers, researchers gave the participants bogus feedback
about how well they'd slept the previous night.[11] Participants
who were told they'd had an above-average amount of REM
sleep performed better on cognitive tests and those who thought
they'd had below-average sleep quality performed worse. How
you *actually* slept, in other words, may be less important than
how well you *believe* you slept, for cognitive-related performance anyway.

With all the anticipation, nervousness, and possibly unfamiliar surroundings, it's almost a given that an athlete won't
have ideal sleep the night before a big competition. That's okay,
Bender says. It's more important to have good sleep the week
leading up to a competition than the single night before an
event. "If you get good sleep the four or five previous nights,
then one bad sleep the night before won't impact you so much."
Ideally, athletes should aim for good sleep every night, but

realistically this isn't always possible. Cheri Mah is a sleep researcher at the Human Performance Center at the University of California–San Francisco. "I don't want an athlete freaking out if she gets five hours the night before a game, but it's false to say that this won't affect her that day." It's clear that sleep deprivation can reduce reaction times and mental skills, in particular. "So many athletes sacrifice sleep or they overlook it. But they prep so carefully for every other aspect of training, and sleep should be held to the same standard."

———

The year 2016 marked Sue Bird's fifteenth season playing in the WNBA, and at age 35, she posted one of her best seasons ever—leading the league in assists and shooting 44.4 percent from three-point range, a career best. Bird has been playing guard for the Seattle Storm since 2002, when she was the first overall pick in the WNBA draft, and in 2016 she started in every single game Seattle played. She also finished the season feeling far less exhausted than in previous years, and for that she credits then-Storm head coach, Jenny Boucek. (In late 2017, Boucek moved to the NBA, as an assistant coach for player development with the Sacramento Kings. Then, in 2018, the Dallas Mavericks hired her as an assistant coach.)

"She's obsessed with sleep. It's almost comical," Bird says of Boucek. "I don't know how she does it, but she sleeps like 12 hours per night, and she plans our entire season around sleep." Boucek's preoccupation with slumber arose from her impulse to question the status quo. "I like to understand the science of the game," says Boucek, a former standout player at the University of Virginia. "I come from a family of doctors, and that gave me the mentality to experiment. I'm not going to do things a certain way just because that's how they've always been done. It's kind of how I'm wired. Our leaders and owners are really bright thinkers and they've given me room to experiment."

The Storm's sports performance consultant Susan Borchardt came from Stanford, where she'd worked with the sleep expert Cheri Mah and had seen the amazing benefits that sleep extension had bestowed on the athletes in Mah's sleep studies. Mah recruited players from Stanford's basketball team and convinced them to sleep or at least rest in bed for ten hours every night for about six weeks (the players ended up getting an average of about 80 minutes of extra sleep per night). Participants in the study improved their reaction times and experienced a 9 percent increase in free-throw shooting percentage and in three-point field goal shooting percentage.[12] Their sprint times improved, to boot. And the benefits didn't stop there. Players who'd extended their sleep also reported feeling happier and more alert. When Mah enlisted the school's swimmers and football and tennis players in similar studies, the results were correspondingly positive. Improve an athlete's sleep, and everything else gets better too.

That was enough to convince Boucek, who realized that some tweaks to the team's practice and travel schedules could help them recover better and fight fatigue. "Sleep is one of those things we all know we should do, but it's hard to get right," Boucek says.

Most basketball teams hold morning "shoot-arounds" on game day, in which players throw the ball around in an informal practice session. Los Angeles Lakers coach Bill Sharman instituted the practice in the early 1970s as a way to help players burn off nervous energy and feel sharp leading into the game, and the practice soon became standard around the league. (Although legend has it that Sharman began the practice as a strategy to roust Wilt Chamberlain out of bed, he told the *New York Times* that the ritual actually began as a way to calm his own nerves during his days as a player.)[13] It's a long tradition, but it can force players out of bed before they're done sleeping, particularly if the team played or traveled the previous evening. Boucek realized that morning shoot-arounds were potentially coming at the expense of her players' sleep. So she cut some of them out entirely and made the remaining ones optional. "Game days are sacred for players, and it should not be up to me to tell all my players what is going to work for them," Boucek says. "Players always have the option of going, and we have coaches available, but they can do whatever routine or rituals help them get ready for the game. If they're needing extra sleep for recovery or mental clarity, I want them to be able to get it."

The Storm aren't alone in rethinking shoot-arounds. Many NBA coaches have also changed their approach to the ritual. "More and more teams are taking the shoot-arounds away and giving the players time in the morning to sleep," says Aaron Nelson, head trainer for the Phoenix Suns. The San Antonio Spurs, Boston Celtics, Portland Trail Blazers, and Denver Nuggets have also experimented with eliminating the morning shoot-around, and some coaches have moved them closer to game time to allow players to sleep in. "Guys are getting several hours more sleep. It helps with the battle against fatigue," Nelson says. The NBA's brutal eighty-two-game schedule, which often features back-to-backs and consecutive games in multiple time zones, has become notorious for wearing players down. Many of the league's coaches have come around to the idea that sometimes rest is more important than more time on a player's feet. "We've talked to a bunch of sleep people to help with the schedule, and we think about things like, when are we practicing? How long are we practicing? When do we shoot-around? Eliminating some of those things has definitely made a big difference," Nelson says.

After a night game, players may not get back to their home or hotel room until 2 a.m. or even later, and it can be hard to wind down after an event. It takes time to come down after a game—win or lose, the rush of game-night energy can't be shut off like a light switch. To give players a chance to sleep late after those long nights, Boucek revamped the team's travel schedule to avoid early morning flights that would cut into their sleep time. "We don't rush to the next city. If you chop up their sleep it's not as effective. Sleeping on the plane isn't as good as sleeping in bed through the night."

The three-hour time difference from the West Coast to the East Coast can disrupt the players' body clocks, and for this problem Boucek came up with a novel solution—remain on your home time schedule. "When we go east, we don't change time zones. We just stay on Seattle time. There's usually nothing on our agenda that would be out of the range of normal if we were still on West Coast time," Boucek says. If players get in and their bodies can't shut down because they're still on West Coast time, that's okay, because they won't have to get up first thing East Coast time. Keeping their clocks on Seattle time prevents players from the stress of not being able to fall asleep while knowing they have to get up extra early, Boucek says.

These changes to the Storm's schedule have made a huge difference on the court. "Players are in better spirits—they're more refreshed," Boucek says. "We don't lose games as much for physical reasons. We're less dragging and brain fried toward the end of the season, where you really see it pay off." The alterations Boucek made to the team's travel and practice schedules to accommodate more sleep are particularly noticeable toward the end of the season, Bird says. "This is my fifteenth year, and I know what a WNBA season feels like. But when she made these changes, I noticed that I didn't feel as tired at the end of the season, and it's because we were sleeping more."

Boucek's penchant for sleep has rubbed off on Bird, who's not only made sleep sacred, but also become a fan of "coffee naps." She drinks a cup of coffee, then lies down for a 20- or 30-minute nap, which is approximately how long it takes the caffeine to fully kick in. "When you wake up, you're ready to go, and you don't have that grogginess."

———

By the tender age of twenty-two, the skier Mikaela Shiffrin had already notched forty-two World Cup victories, won two Olympic medals (gold and silver), and secured her second win in the World Cup overall.[14] If she continues on this trajectory, she may well become one of the greatest skiers of all time. Despite her rapid success, Shiffrin lives a monk-like lifestyle, eschewing celebrity events and late-night parties in favor of healthy meals and an early bedtime. She credits her parents with instilling her good habits. "My mom's a nurse and my dad's a doctor and they have always understood the importance of nutrition and sleep. They taught me that at an early age," she says.

Napping is Shiffrin's number one recovery strategy. "I took one of those BuzzFeed quizzes and it said that my spirit animal is a sloth and my nickname is Sir Naps A Lot. It's pretty accurate, actually. I love to take naps. I put aside a solid two hours every day to make sure I have that time to rest. When you're sleeping you're recovering physically, mentally, and emotionally." Her napping habits have become contagious. At a recent training camp, she religiously stuck to her afternoon nap schedule. At first, she recalls, other skiers would say, "Oh, I don't take naps." But as the intensity and amount of training rose during the camp, other athletes slowly came around to her ritual. "They'd say, well, maybe I'll just lie down for a bit too. We'd all sort of have our naptime."

Sleep is so important to Shiffrin that she always travels with a comfortable pillow and what she calls "the softest blanket I have ever felt," which she received as a Christmas gift. The blanket has become such a cue that she takes it everywhere—she's afraid she won't be able to sleep without it. Her naps are as much a part of her training routine as her ski runs. The first thing she does when she gets to a new hotel room is look at the curtains. "I'm very excited when it looks like I can black out the room. If the curtains aren't great, I'll wear an eye mask or cover the windows with towels."

While Shiffrin has heard some people say that napping prevents them from sleeping well at night, her philosophy is to get sleep whenever she can. She usually sleeps for eight-and-a-half to nine hours every night. "If I had a big nap that day, then I'm fine if I don't get a solid nine hours that night. I can't function at the level that I want to if I don't get enough sleep." On the rare occasions where she only gets seven hours of sleep, she wakes up feeling like she's in a daze. "I don't even feel comfortable driving," she says, adding that she's perplexed by people who brag that they're fine on five hours.

Olympic gold medalist Lindsey Vonn is also known for napping.[15] Naps are so important to the US ski team that their training center in Park City, Utah, includes a sleep center where athletes can snooze between training sessions.[16] The nap rooms have acoustic ceilings to reduce distracting noise, black-out shades to block light, and temperature controls to keep the conditions perfectly cool for sleeping.

Napping was once a ritual reserved for kindergarteners, but it's become a common habit among elite athletes. Some NFL teams also have nap rooms, and afternoon napping has become so prevalent in the NBA that "Everyone in the league office knows not to call players at 3 p.m. It's the player nap," league commissioner Adam Silver told the *New York Times*.[17] "If you nap every game day, all those hours add up and it allows you to get through the season better," LA Lakers point guard Steve Nash told the *Times*. LeBron James, Derrick Rose, and Kobe Bryant are known to partake of pregame naps.

"Naps are really important for athletes," says Amy Bender, the Canadian sleep scientist. "We know that naps improve alertness, they improve motor performance, and they improve productivity." The ideal time for naps is during the body's natural dip in the afternoon, she says, sometime between 1 and 4 p.m., and they should be scheduled in as a part of the normal training routine.

Naps can also help bank some sleep in advance of nights when snoozing might not come easily. Competition or hard training can interfere with sleep, and fatigue does not always equate with sleepiness, Australian Institute of Sport recovery researcher Shona Halson says. Studies suggest that athletes don't always sleep easily during periods of hard training, where they may report feeling "overtired."

At the 2016 Olympics in Rio, the swimming events went late into the night. "Some of the finals went at 10 p.m. at night, and the head coach came to us and asked, 'How are our guys going to cope with this?' " says Halson. "No one had ever looked at this," she recalls, so they decided to do some experimenting. The AIS scientists got the team together and made them do some time trials at midnight, then they monitored their sleep to find out how they coped. The experience provided the athletes a chance to find out "what it feels like to swim fast at midnight, and it gave them the confidence to know that they could perform at that time," Halson says.

If you want to perform at a certain hour, then ideally, that's when you should train. "It's training the sleep muscle to do what you want at the time that you want," says Meeta Singh, the scientist at the Henry Ford Sleep Disorders and Research Center. She advises athletes to train their body clocks so they're accustomed to being awake and asleep at the times they'll need around competitions. "If your games are always going to be at 7:00 p.m. then that's when you should do your training. The basic advice is always to train the time when you know you're going to play."

Night competitions can be challenging enough. Pair them with jet lag and the problem is even worse. In this respect, West Coast teams like the Storm have an advantage. One analysis of West Coast versus East Coast matchups in the NFL over forty seasons found that teams coming from the West have a significant circadian advantage in evening games—West Coast teams beat the point spread about twice as often as East Coast teams.[18] How an athlete responds to time changes depends a little bit on their natural chronotype—whether they're a night owl or an early bird. If you're a night owl, your circadian rhythm is likely longer than 24 hours, and this makes it easier for you to fly west than to travel east, Bender says. "For morning types, it's easier to fly east, because their circadian phase is shorter."

When traveling to a new time zone, Bender recommends

that athletes use light to help their bodies adjust. When heading east, blocking out light, especially blue light, at night and then seeking morning light can help shift the body clock earlier. Heading west, the opposite holds true. You want exposure to evening light and to avoid morning light so that you can delay your body clock to sync with the new time zone, Bender says. These days apps and websites like Jetlagrooster.com can give individualized advice on how to use light and shifting bedtimes to adjust to time zone changes.

The best way to overcome challenges like jet lag is by planning ahead and adopting good sleep habits as part of one's daily routine, not just in advance of important events. Athletes tend to give the most thought to sleep immediately before competition, and they care less about habitual good sleep, but that's a mistake, says Shona Halson. Like everyone else, athletes have FOMO—fear of missing out. Halson recalls one athlete she worked with who seemed off his game. She finally found out why—he'd been getting up in the middle of the night to check the final standings in overseas cricket games. "He couldn't wait a few hours to get the scores," she says.

There's a reason so many teams and performance centers are starting to provide nap rooms for their athletes. Sleep is the number one thing that athletes can do to bounce back from their training. It's like the cake of recovery. Everything else is just icing.

8

Selling Snake Oil

On a spring day in 2017, I paid a visit to a GNC store in an upscale Palo Alto strip mall to find out what they could offer me for recovery. The shopkeeper greeted me with a smile, and when I asked what he had for recovery, he pulled a giant tub of protein powder down from a shelf. For $55 (a little less than $4 per serving) GNC Pro Performance® Amp Amplified Wheybolic Extreme 60™ Original promised to fuel "a 30% increase in muscle strength" with its "amino acceleration system." This claim was followed by a pesky asterisk, pointing to a disclaimer: "These statements have not been evaluated by the Food and Drug Administration."

The package did point to some science—an 8-week study that supposedly showed that athletes taking the stuff while performing an intense resistance exercise training regimen made greater gains in muscle strength and size than those on a placebo. The label also claimed that an 8-week study of 30 healthy male volunteers found that the men taking the supplement showed equal gains in maximal muscle strength and muscle endurance compared to the control group. It sounds promising, but what does that really tell us? It's hard to assess studies based on these single-sentence descriptions.

The shopkeeper wasn't much help. He told me that the store's owner swore by this stuff, but he didn't know anything about the studies referenced on the label. The labels tout science, but the sales seemed to rely on personal anecdotes and success stories from the shop owner or some guy next door. The tub also said it contained the "highest-quality, fast-absorbing forms of whey protein—isolates and hydrolysates," but when I asked him where and how the ingredients were manufactured, he admitted that he didn't know.

Back at home, I went on the GNC website looking for more details about the studies referenced on the label, but still didn't find enough information to be able to look up the studies for myself. What I did find was lots of products for recovery—forty-four in all. The claims were mind-boggling. One purported to "work synergistically with the body's own mechanisms of renewal" and touted an enzyme blend supposed to promote "joint comfort and healthy circulation" and to encourage "accelerated muscle and tissue recovery." Every claim came with an asterisk to indicate that it was unsubstantiated by any outside agency. Another recovery supplement was said to contain a "scientifically engineered formula" that has "pinpoint accuracy on the body's reservoirs to build and replenish." Whatever that means.

If the trademarked proprietary ingredients seemed too engineered, I could turn to an equally dubious array of "natural" products. On a recent trip to my local health food store, I found an entire display shelf packed with recovery products— everything from an "organic recovery nectar" (made of freeze-dried coconut water) that promised to "decrease recovery time and soreness" and "help repair torn muscles," to a watermelon-flavored drink mix that claimed to "optimize muscle recovery" and "reduce fatigue and muscle soreness."

How legit are these promises? "Pretty infrequently do these things live up to the claims," says Anthony Roberts, who spent years working inside the supplement industry (and the steroid industry prior to that) before he turned to writing critically about these issues. "There's a lot of misinformation out there, and it's generally funded by people with a vested interest."

Expert groups who've evaluated supplement claims have reached similar conclusions. According to a 2016 consensus statement released by the Academy of Nutrition and Dietetics, Dietitians of Canada, and the American College of Sports Medicine, "few supplements that claim performance-enhancing benefits have sound evidence to back them." Studies touting sports supplements are often limited by small sample sizes and participants who aren't representative of the athlete population. It's my beer study all over again.

———

During the introductory nutrition course he teaches at McMaster University, Stuart Phillips likes to ask students if any of them take a multivitamin. A lot of hands go up, he says, and when he asks, "*Why* do you take it?" their answer is almost always the same: "to make up for what I might be missing." This FOMO (fear of missing out) drives a lot of the marketing around sports nutrition products, particularly supplements. "They really hammer home the message that 'you might be missing something!'"

If your overall diet is pretty good, you don't need to worry about getting special vitamins or nutrients after exercise. "But that's where a lot of the marketing comes in, because most people are under the impression that their nutrition is poor," Phillips says. Supplements are marketed to exploit this anxiety. According to the Centers for Disease Control and Prevention, the typical American diet—and we're not talking about health nuts here—appears perfectly adequate for getting the nutrients and vitamins needed for good health.[1]

Of course, the CDC reports are just referring to the minimum needed for good health, and athletes don't just take supplements out of worry that they're missing some vitamin. They want to become superhuman. The hope is that they'll optimize their body's inner workings and give them some performance boost—like doping, but legal.

A 2017 report from the International Olympic Committee concluded that "there is a limited evidence base for most supplements." What drives most supplement use, the report said, was "the fear that an athlete cannot afford to miss out on what their rivals are using." What makes supplements so popular among athletes isn't that the products are so incredibly helpful that they can't live without them. In all of my reporting, I have never found an athlete who pointed to a supplement pill or powder having led to an amazing breakthrough. (Some of them *have* said that a supplement changed their lives forever, but we'll get to that in a moment.) Instead, it's that when they look around, it seems like everyone's doing it. No one wants to be left out—FOMO is a powerful thing.

———

The first time I met Mary Beth Prodromides, I recognized her from a block away. I knew her only from an article I'd read in the newspaper, and it wasn't so much her appearance that tipped me off, but the way she glided across the street with the poise of a superhero. Her steely shoulders and chiseled quadriceps made her appear much more imposing than her 5 feet 2 inches and 130 pounds.

Prodromides throws around barbells and weight stacks like they're made of bamboo, and four times now, the CrossFit Games has crowned her the world's fittest woman in her age class, currently 55 to 59. Her hard body is a contrast to her quiet manner, and she has a gentleness about her—at least until it's time to throw down. The tanned brunette often wears her long hair in a ponytail while doing clean and jerks or squats, which she performs loaded with enough weights to clean out most gyms. Prodromides has a long history as an athlete. She's been a gymnast and a bodybuilder, but in CrossFit, she may have found her perfect sport.

The CrossFit Games bills itself as the world championships of "the sport of fitness," and it combines Olympic weight lifting exercises with gymnastics moves and some cardio like running or rowing on a stationary machine. Securing an invitation to the Games requires excelling at a complex maze of qualifying events, and winning the competition takes the ability to perform a random sequence of exercises (many of them not announced until just before start time) with speed. "I am good at exercising fast. If you had told me when I was 16 that I was going to be a really great fast exerciser I would have never believed you. But that's what it takes to be good at CrossFit—go fast and not get hurt," Prodromides told me.

Regular CrossFit goers show up at their box and do the posted workout of the day, or WOD. But like other top CrossFit Games competitors, Prodromides, a P.E. teacher at a middle school in Grand Junction, Colorado, trains more like an elite athlete, with a carefully programmed and periodized training schedule overseen by a professional coach. In the lead-up to the Games, she spends around four hours per day in the gym, and recovery is crucial. For help, she turns to one of her sponsors, AdvoCare, a multilevel-marketing company that sells a wide variety of vitamins, supplements, and drink powders.

Prodromides's supplement regimen and endorsement deal are hardly unique. The sports supplement industry is expected to reach a value of $12 billion annually by 2020.[2] Much of the money the supplement industry spends promoting its wares goes to professional athletes, teams, leagues, and athletic events. AdvoCare alone has sponsorship deals with NASCAR, the MLS, and multiple NCAA events, as well as with individual athletes like NFL quarterback Drew Brees. Numerous other multilevel-marketing and supplement companies have struck deals with high-profile sporting events and teams. Herbalife sponsors the LA Galaxy Major League Soccer team and some Ironman triathlon events and Usana Health Sciences sponsors the Women's Tennis Association and the US ski and snowboard teams.[3] The US national gymnastics, rugby, triathlon, rowing, pentathlon, fencing, taekwondo, softball, and soccer teams have all inked sponsorship arrangements with Thorne Research, an Idaho-based maker of sports supplements, and Garden of Life organic supplements is an official sponsor of USA Track & Field. Social media is cluttered with well-known athletes posting photos and glowing praise of their sponsors' supplements, and these ubiquitous pitches and paid endorsements lend an aura of legitimacy to these products and create a sense that everyone's using them.

Before our second meeting, I asked Prodromides if she'd show me what she takes. She met me at a coffee shop, arriving with a plastic shopping bag full of products that soon filled the small table between us. "The first thing is, supplements don't help you if you don't have good nutrition in the first place," she said, by way of caution. "You're not going to fix anything. Your supplements can't be your food. That doesn't work."

Prodromides began using AdvoCare in 2010, shortly after she started doing CrossFit. You can't buy the company's products in your local vitamin store. Instead, its wares are sold through an affiliate system. At the bottom level are people who pay retail to purchase products from a distributor. If you sign up to become a distributor, you get a price break on your products and a cut on everything you sell. The next level up, advisors, can get bonuses for signing up distributors to work under them. The company promises that people who work hard to build their AdvoCare business can reap life-changing financial rewards, but a 2016 investigation by ESPN reporter Mina Kimes found that, as with many multilevel-marketing programs, very few people make even a modest income from selling the products.[4] According to Kimes's report, participants in the system are pressured to buy inventory to keep their status as distributors, and anyone who questions the AdvoCare model is labeled a "dream killer."

A guy at her gym at the time introduced Prodromides to AdvoCare, and, like many users, she began the program with AdvoCare's 24-day challenge, which is touted as the jump start your body needs to reach your goals. Prodromides had just started doing CrossFit and she was changing her diet too—no processed food, no wheat, very little dairy. The twenty-four-day challenge gave her something of a fresh start.

She finished the challenge and soon became an AdvoCare advisor, which means that in exchange for meeting a certain threshold of sales, she gets 40 percent off the retail price of the company's products. She doesn't put a lot of effort into selling the stuff—she's too soft-spoken and earnest to give anyone the hard sell—and she never tried to make a sale with me in all the times we spoke. I believe her when she says that she's not into pushing the stuff on anyone. "If people ask me about it, I'll tell them. But I'm not really trying to get people under me," she says. Now that she's a sponsored AdvoCare athlete, which she negotiated after her first win at the CrossFit Games, she gets a monthly stipend, paid in product. Although she no longer has to pay for her own supplements, she remains in the advisor program and makes a little money from that too. "It's not much, like $100 per month, if that. But it helps."

At retail prices, a 30-day supply of her regimen would cost more than $730. Even with the advisor discount, the price would top $372 each month. Her program, which she laid out for me on the table, consists of twelve items: Omegaplex fish oils, Bio-Tune, BioCharge branched-chain amino acids, Immunoguard, Joint ProMotion, calcium, V16 Energy, SPARK energy drink, VO2 prime energy bars, Nighttime recovery, Sleep Works, and a postworkout recovery shake.

"The omegas and the postworkout recovery shake are the ones I never miss," she says. What does she like about the recovery shake? "It's chocolate!" She mostly adheres to a paleo diet that avoids grains and processed foods in favor of meat, fish, vegetables, nuts, and vegetable oils, but "I'm not strict paleo because of what's in this shake. That's my one cheat." The AdvoCare postworkout sports performance drink mix contains more than fifty ingredients, beginning with two sugars—maltodextrin and fructose—and continuing with a long list of chemical names like bromelain, gamma-oryzanol, and choline dihydrogen citrate. The ingredients also include creatine—a compound purported to improve muscle strength—and a string of vitamins and minerals. According to AdvoCare, the shake "helps minimize occasional muscle soreness, optimizes muscle recovery, supports muscle metabolic processes, and helps maintain and restore energy supplies after physical activity." The company provides no evidence to support these claims.

On the other hand, there's pretty good evidence that the carbs and protein in the shake can help an athlete like Prodromides refuel and recover after a hard workout. One serving contains 36 grams of carbohydrates, 12 grams of protein, and 220 calories. She could get approximately the same nutritional value from a bowl of Greek yogurt and a banana, but the shake mix is convenient, and Prodromides likes the taste.

As for the other products, she says the BioTune helps with "the stress of exercise and the stress of life," the Immunoguard prevents her from catching all the bugs that she's exposed to from the kids at her school, the Sleep Works helps her sleep better, and the Nighttime recovery helps her body repair. "It's different from the others, because it tells you how much to take based on your weight," she says. If the dose varies like that, it must be powerful. She takes the BioCharge branched chain amino acids because "it's what Drew Brees uses" (even elite athletes aren't immune from celebrity endorsements), and the Joint ProMotion and calcium help her bones and joints. She doesn't take a VO2 prime energy bar every day, and when she does, she often eats only a few bites. The V16 Energy is a vitamin and herbal powder that's poured into water to make an effervescent drink that AdvoCare claims can support energy and mental focus (while noting that this claim has not been evaluated by the FDA). The Spark energy drink contains caffeine and a bunch of vitamins. Prodromides says she can't usually drink a whole packet. "It gets me too pumped up." I wonder aloud how she can know which supplement is doing what and which ones are really working when she's taking so many things at once. She explains that if she doesn't think something is working she'll stop taking it, but I'm not sure that answers the question.

A serious athlete like Prodromides can't leave any potential advantage untapped, and AdvoCare makes some pretty enticing claims. The BioCharge supplement, for instance, promises to "improve muscle performance and recovery through a formula of antioxidant rich botanicals combined with B vitamins and branched-chain amino acids (BCAAs)." The product also promises that the BCAAs "contribute to reducing muscle damage and soreness" and help "combat the stress of aging and the stress you put on your body when you workout or compete." Amazing, if true, but is it? Numerous studies of BCAAs have been published, but most are limited in scope, and any positive effects have been small. As with so many supplement studies, the sample sizes are small. Like my beer study, they're interesting, but far from definitive.

———

If anyone knows the performance-enhancing promises made by supplement companies, it's Jose Antonio. Based in Florida, Antonio is an avid stand-up paddleboarder, and he's also CEO and cofounder of the International Society of Sports Nutrition and an editor for the society's journal. He was previously the science editor at *Muscle and Fitness* magazine, has worked for numerous companies in the supplement industry, written multiple books about supplements, and has built a reputation as a supplement guru. "Sports nutrition science is the black sheep of the academic family," Antonio told me. Twenty years ago, supplements were mostly shunned by the research community. "People in exercise science treated it like snake oil," he says.

But Antonio and his associates in the supplement industry realized that they didn't have to give academic sports scientists the final word. Instead of trying to convince the naysayers, they could invent their own field of research, and if they formed their own research society and published their own academic journal, they could build legitimacy. Antonio and four other supplement enthusiasts founded the ISSN in 2003, shortly after Antonio and another cofounder, Jeff Stout, gave a seminar at a meeting of the American College of Sports Medicine. Their lecture drew criticism from members of the audience who said that they were peddling quackery and there wasn't any evidence for the supplements they were touting. But even as half of the attendees attacked their message, the other half "loved it," Antonio recalls.

Researchers who belonged to the ACSM were mostly academics and sports physicians. Antonio and his colleagues came from a different world: bodybuilding. "All of us who started studying it scientifically, our interests actually came from bodybuilding—either we were fans or we competed. And we knew, based on what we did to our bodies, that they were wrong," Antonio says. "Doctors were telling us you don't need all this protein if you body build, yet we were body-building and we knew we needed protein. Clearly there was something wrong."

Antonio and his fellow ISSN founders believed that very few people possessed genuine expertise in sports nutrition, and they saw a need for a group that would support their interests and knowledge in the subject. "We were rebels. So we said sayonara to the traditional academic societies and formed the ISSN," Antonio said in a 2014 interview.[5] "Let's face it: the

ISSN is like a speed boat. The other academic societies are like paddling a dinghy across the Hudson Bay." Antonio and Stout were working for the supplement companies GNC and MET-Rx around the time they founded ISSN. "We were called sellouts, snake-oil salesmen, you name it. But we got the last laugh. Now you'll find sports nutrition as an academic field of study."

Antonio credits the ISSN with legitimizing the field of sports nutrition and the use of sports supplements. Prior to about 2000, journals didn't want to publish anything that showed that supplements might help, he tells me. The ISSN created its own journal to provide an outlet where people like him could publish supplement studies. "Now, there's such a growing demand for information on supplements that we're not the only journal who does it—because it's cool, it's sexy, it's fun and people love it!" The ISSN may be like a speed boat, and Antonio is more like a cruise ship's enthusiastic activities director.

Supplements got a bad reputation, Antonio says, because they originally came from the world of bodybuilding. "If you understand the culture of bodybuilding—most of the marketing claims were just made up, and we all knew they were made up. The funny part is they actually got it right half the time." In the early days of bodybuilding, he says, muscleheads like Joe Weider and Jack LaLanne talked about supplementing their food with extra protein. "Scientists and doctors thought they were wrong, but it turned out they were right."

Maybe, but it's also true that many of bodybuilding's early and best-known figures had financial reasons for backing supplements. A look back at the history of bodybuilding and supplements shows that the relationship wasn't just about gymgoers stumbling upon muscle-building magic.

Bob Hoffman was an influential name in the bodybuilding

world for more than three decades, beginning in the 1950s.[6] He coached the US weight-lifting team for many years, owned York Barbell, and founded the magazines *Muscular Development* and *Strength and Health*. In June 1946, natural food advocate Paul Bragg wrote a letter to Hoffman encouraging him to consider selling food products.[7] "We can really add a tremendous income to your earnings, because the food business is not like the athletic equipment business. In 1913 I bought a set of barbells from the Milo Barbell Company and today they are just as fine as they were way back there in the dim past. But when you get thousands of your students eating your food and they consume it, you have no idea of the tremendous income that you will have rolling in." Hoffman wasn't immediately convinced, but by 1951, *Strength and Health* was featuring a half-page ad for "Johnson's Hi Protein Food," which was "endorsed and recommended by Bob Hoffman, Famous Olympic Coach" and available exclusively from the York Barbell Company. Hoffman eventually became known for selling a rebranded, reformulated version of the product: "Hoffman's High-Proteen."

In light of this history, I expected Antonio to name a long list of supplements when I asked what he recommends for recovery. Instead, his advice was this: "One, sleep. There's a lot of data that you just need to sleep a lot." He also suggested 20 to 40 grams of protein postexercise, and some carbohydrates to replenish muscle glycogen if you're an endurance athlete. "If you stick to the basics you'll do fine. The problem is, most people can't even do the basics. People are like, hey, what's the secret? I'm like, well . . . you train really hard, you sleep a lot, you eat well, and you repeat it a lot."

It's good advice, and Antonio clearly understands some things about rigorous science. "One study doesn't prove anything. One study is just part of a larger body of work. You have to look at the entire body of work," he says.

Most supplements, though, have very little evidence behind them. To that, Antonio essentially says, so what? Lots of claims about food aren't proven either. "There are people on the other side, mostly dietitians, who think that food does everything. And I always challenge them and say, show me, where are the food studies? Where are the studies showing that eating a ham sandwich postexercise helps? Where's the study showing that ketchup is safe?"

If a supplement manufacturer makes a claim, ask them for the published science on their product, Antonio says. "If they can't provide that, it doesn't mean they've never done a study. It doesn't mean the product claim is wrong, but it means they haven't taken the time to do their due diligence to actually fund a study, which I think supplement companies should do. There's no better marketing than science."

No better marketing than science. . . . that's the problem with many supplement studies—they're not scientific quests for truth, but marketing exercises designed to sell products.

The same criticism has been lobbed at trials sponsored by pharmaceutical companies, which is why clinical studies must increasingly follow stringent rules designed to make drug studies more transparent. Researchers running drug studies in the United States are now required to register their plans and protocols in advance at ClinicalTrials.gov and to report and share *all* their results, not just the positive findings.[8] These requirements were implemented to ensure that regulators and the public get the full picture, and not just results cherry-picked to make a product look good. The system doesn't work perfectly and it can't cull every false positive finding, but a 2015 analysis showed that after the launch of ClinicalTrials.gov in 2000, the proportion of trials that found a positive benefit plummeted.[9] Prior to 2000, for example, 57 percent of large randomized, controlled trials sponsored by the National Heart, Lung, and Blood Institute showed that the treatment worked. After 2000, that number dropped to 8 percent.

But no such rules or conventions apply to nutritional studies, which means that supplement companies and the researchers they fund can fish around for positive results until they find something that paints the product in a positive light. Null results (studies that don't show any effect from the supplement), meanwhile, can be hidden away in a file drawer. To be fair, sometimes this so-called file drawer problem happens not because researchers are doing anything nefarious, but because journals favor positive findings over negative ones and novel findings over replications. It's often easier to get positive findings published than negative ones. But the end result is the same: the published scientific literature can end up weighted toward positive results.

Most journals have a peer review process that's meant to weed out studies that are too flawed to provide meaningful information. But the competition to publish in top-tier journals is fierce, and if you've got a study to publish, you can now bypass peer review altogether by submitting it to a journal that looks legit, but doesn't subject papers to peer review or rigorous scientific standards. These so-called predatory publishers—journals that, for a fee, will publish nearly anything—have flourished in recent years. (One accepted for publication a paper titled "Get Me Off Your Fucking Mailing List," whose text was nothing more than those seven words, repeated over and over for ten pages.)

In a perfect world, science is supposed to seek truth and go wherever the evidence leads. But in practice, studies may be used as marketing tools by designing them to back claims that a company wants to make about its product. In that case, you put together a study, perhaps tilt the scales in their favor with a design that helps to produce the desired outcome, and bam— you've got your "scientific proof." You don't even need the study to pass peer review as long as you're willing to pay a bit of money to publish it. And if the study doesn't go as planned, you can just file it away somewhere that no one will see it and try again.

For five years, University of Colorado librarian Jeffrey Beall ran a blog listing predatory publishers. After shuttering the blog in early 2017, he published a paper sharing what he'd learned.[10] "I think predatory publishers pose the biggest threat to science since the Inquisition," he wrote. "They threaten research by failing to demarcate authentic science from methodologically unsound science, by allowing for counterfeit science, such as complementary and alternative medicine, to parade as if it were authentic science, and by enabling the publication of activist science." Distinguishing science from marketing has become a little bit like weeding out fake news on the internet. Established, legitimate sources are easy to spot, but then there's the ever-expanding mass of next-tier sources, some of which are credible, and some of which are not.

———

Most supplement-takers probably understand that the claims may be inflated. I've talked with many athletes who say that even if they're only half as magical as promised, at least it's something, and the worst-case scenario is that you're spending money on something that doesn't work.

Triathlete Lauren Barnett used to think that way too. Then

she was notified that the drug test she'd taken after her win at
the 2016 half-Ironman triathlon in Racine, Wisconsin, had
come up positive for ostarine, a banned muscle-building drug
she'd never heard of. "I was unaware of even how to pronounce
it," she says. She'd never knowingly doped, and had no idea how
this unfamiliar substance could have ended up in her body. Eventually, after testing (at her own expense) nearly everything she'd
come into contact with before the race, she found the culprit—
an electrolyte tablet she'd taken to help her cope with Wisconsin's July heat. (If only she'd realized that electrolytes are just
salts that she could have gotten from her breakfast food.) Anti-
doping officials independently purchased the salt tablets Barnett had taken and because their tests confirmed the presence
of ostarine, Barnett received a reduced six-month ban.

Barnett was hardly the first athlete to ingest an illicit drug via
a tainted supplement. The attorney who represented her, Howard Jacobs, has made a career out of defending athletes who've
tested positive for doping tests after consuming banned substances unknowingly, through a tainted supplement. (He's also
defended athletes who later confessed to cheating—like my old
University of Colorado cycling teammate Tyler Hamilton, who
knowingly doped yet insisted he hadn't.)[11] Other Jacobs clients
include swimmer Kicker Vencill, who in 2003 tested positive
for a banned steroid precursor, which he traced back to something seemingly innocuous: his multivitamin. He missed two
years of competition and a chance at making the 2004 Olympic team. Vencill sued the vitamin's maker, Ultimate Nutrition,
and won a nearly $600,000 settlement.[12] "It should be a wake-
up call," says Jacobs.

Jacobs also defended swimmer Jessica Hardy, a world record
holder who missed the 2008 Summer Olympics after test-

ing positive for clenbuterol—an asthma medication that can increase muscle growth—at the Olympic trials. An independent lab tested her supplements and traced the drug back to a product called Arginine Extreme, made by the same company—AdvoCare—that sponsors CrossFit champ Prodromides. The company denied culpability.[13] But the evidence Hardy presented in her hearing was compelling enough to convince the World Anti-Doping Agency that she'd inadvertently doped via a contaminated supplement, and she received only a one-year suspension instead of the usual two-year ban.[14]

The CrossFit Games also does drug testing, but Prodromides is not worried. AdvoCare has joined a program called Informed Choice that tests products to look for substances on the antidoping banned drug lists. "It guarantees that what's in the product is on the label, so I don't have to worry about drug testing. I know it's legit," Prodromides says.

Jacobs is just one of a half-dozen experts who told me that athletes are foolish to bet on such assurances. "Any athlete who thinks, I've done my research and it's impossible this product could be contaminated, is kidding themselves," he says. Although numerous programs have sprung up with promises to ensure supplement quality, "it's not a guarantee, and you know that it's not a guarantee, because none of them are offering big payments to athletes if they happen to cause a positive test." He says that no company can guarantee that their products are pure and free from contamination.

On the surface, the supplement excuse seems like a convenient cover. Dopers can find some dubious supplement and then use it as a shield if they get caught. There's no doubt this approach has been tried. In 2012, San Francisco Giants star Melky Cabrera went so far as to create a fictitious company and

a fake supplement that he could finger for the positive testosterone test that was set to land him a fifty-game suspension.[15]
Instead of exonerating Cabrera, the stunt, which involved paying $10,000 to create a phony website for the fake product, left a
trail of clues that led back to Cabrera himself.

Athletes who fail a drug test can't just blame a supplement
to get a reduced sanction, says Amy Eichner, special advisor
on drugs and supplements for the US Anti-Doping Agency. The
agency needs to independently verify the contamination before
they'll consider any deals. Her agency has spent years desper-
ately trying to educate athletes on the dangers of taking supplements, which can be spiked or contaminated with steroids,
stimulants, or other illicit and potentially dangerous ingredients that aren't listed on the label. It's a problem that will only
get worse as the sports supplement industry grows.[16]

The problem of contaminated supplements has become so
urgent that USADA created a program called Supplement 411
to inform athletes about the risks of ingesting a banned drug or
potentially dangerous substance through a supplement.[17] The
group cannot and will not assure athletes that any supplement
is safe, but USADA does keep a long warning list of products
shown to contain unlisted or risky ingredients. "We routinely
find prohibited substances in supplements we test," Eichner
told me. It's not exactly a new problem—the IOC Medical Commission has warned athletes of the potential risks posed by supplements since 1997.[18] Eichner and coauthor Travis Tygart, CEO
of the US Anti-Doping Agency, estimate that there are more
than a thousand products out there that contain undeclared
stimulants, anabolic-androgenic steroids, or pharmaceuticals.

How do undeclared ingredients get into the supplements in
the first place? There are two likely scenarios. One, the ingredi-

ent has been put there deliberately to make the supplement do *something*, preferably the thing that it's purported to do. This is why supplements for erectile dysfunction, which, like any supplements, are not supposed to contain prescription drugs, are often laced with Viagra, and weight-loss pills commonly contain undeclared stimulants. Many of the FDA's warnings about sport supplements have involved products with hyped-up names that were found to contain anabolic steroids, but even seemingly benign things like protein powders have turned up undisclosed ingredients.

A *Consumer Reports* investigation in 2010 examined protein powders and ready-to-drink protein beverages and found that some of them contained enough arsenic, cadmium, or lead that three servings could cause a consumer to exceed the US Pharmacopeia's (USP) daily limits.[19] At least two athletes, NFL running back Michael Cloud and Olympic bobsledder Pavle Jovanovic, have fingered a protein powder as the source of steroid precursors that turned up in their doping tests.[20] (Their lawsuits against the products' makers were settled out of court.)

Despite assurances by supplement companies that their products are pure, USADA counsels athletes that it cannot ensure the safety or purity of any supplement. Sure, products certified by a program like Informed Choice are probably less likely to be adulterated than those that aren't, but it's buyer beware.

—

Never mind those elite athletes failing drug tests—tainted supplements can also make you sick. A *New England Journal of Medicine* study calculated that dietary supplements are responsible for 23,000 emergency room visits per year in the United States.[21] In 2011, after two soldiers died after taking supplements con-

taining an amphetamine-like stimulant called dimethylamylamine, or DMAA, the US Department of Defense banned the
products from being sold on military bases.[22] In 2013, the FDA
issued a warning that the DMAA in some supplements, one of
them a USPlabs product called "Jack3d," could, well, jack heart
rate and blood pressure and possibly lead to a heart attack.[23] The
supplements remained on store shelves while the agency asked
manufacturers to provide evidence that the ingredient was safe,
but in July 2013, USPlabs voluntarily destroyed its inventory
of DMAA-containing products at its Dallas facility while the
FDA continues to advise people to avoid products containing
DMAA.[24]

Supplements from different companies may seem unique,
but they're often made from the same raw materials, and many
of these raw ingredients are made overseas, often in China.
Where did the amino acids in your protein powder come from?
Good luck finding a good answer, says Anthony Roberts, the former supplement insider who now writes about the industry. The
ingredients in supplements are often sourced from animal by-
products, he says, but it's hard to definitively trace the origin of
most ingredients.

It's not enough for the company packaging and selling the
supplement to be upstanding. Their suppliers have to ensure
the quality of the raw materials, and assurances aren't always
reliable.[25] Nearly every week, it seems, the FDA circulates an
announcement about a supplement tainted with an undeclared
pharmaceutical or causing some kind of medical harm. These
announcements and warning letters don't necessarily put a
stop to the sale of the products, however. Some companies who
receive warning letters continue to sell their risky products.[26]

Why isn't the FDA ensuring that supplements are safe

and unadulterated before they go on sale? Because, by law, it can't.[27] The 1994 Dietary Supplement Health and Education Act (DSHEA) gives the FDA no authority to require supplement manufacturers to demonstrate the safety or effectiveness of their products. Instead, it's up to supplement companies to ensure that their products are safe, and the FDA can only ask for a recall if the agency can prove that a supplement is harmful. The rules make it very difficult for the FDA to remove dangerous products from the market, says Joshua Sharfstein, a former deputy commissioner at the FDA. "Without greater oversight I think it's dangerous to take supplements," Sharfstein says. The FDA had identified nearly eight hundred different brands of supplements that are adulterated with drugs, yet only a small fraction of them have been recalled. Even when the FDA identifies a problem, it can take years to do anything meaningful about it. For example, more than one hundred people, including Baltimore Orioles pitcher Steve Bechler, died after taking ephedra, a supplement that purported to increase energy, and yet it took the FDA ten years to ban it.

Attempts to give the FDA more power to protect consumers have been blocked by the nutrition and supplement industry, which spent $4 million on lobbying efforts in 2014 alone, and contributed another $1.1 million to political candidates and groups.[28] Under DSHEA, the FDA is not responsible for checking the safety of supplements before they are sold, nor is any government agency required to test supplements for safety or efficacy. You might assume that when you see an official-looking product on the shelf that some regulator has inspected it. But it turns out that the FDA may not even know that it exists. The agency doesn't even have a way to track what's being sold.

In a 2009 blog post, Roberts described how he legally intro-

duced a new herb into the supplement market without conducting any safety trials or clinical research.[29] The supplement
contained an herb, *Fadogia agrestis,* which appears to have
been studied primarily in lab animals. "All I had to do was read
a study that said it boosted testosterone, and a bit of googling
later I found out that it had been used for decades in Nigeria as
a folk remedy to treat erectile dysfunction, and I was off and
running. Hell, it's been used for years in Nigeria—it probably
won't kill anybody. But the kicker is that the FDA would have
had to prove that it was unsafe to have it pulled off the market—
not the other way around. I didn't have to prove it was safe at
all," Roberts wrote. He gave an online supplement retailer a
50 percent cut of the royalties in exchange for advertising the
product, dubbed MyoGenX, which quickly became a best seller.
"Unfortunately, after the product was on the market for several
months, a study was published showing that the dose I'd been
using—1500mgs/day—was unsafe," Roberts wrote.

By then, MyoGenX was off the market, but other companies were making copycat products that hid the amount of the
herb actually present in the product by calling it a "proprietary
blend," which allows manufacturers to keep the amount of the
ingredients a secret. The rule was intended to protect intellectual property, Roberts says. "Instead it allows manufacturers to
'pixie dust' their products with ineffective doses, and still get to
claim they have the ingredient on the label." Roberts suspects
that the copycat products didn't contain much of the herb, but
who really knows?

———

If supplements are so dodgy and unproven, why do so many athletes still take them? "We just want to believe in magic," says

Catherine Price, author of *Vitamania*, a history of vitamins and the supplement industry. "We're primed to believe it. The supplement industry has done a great job of convincing us." John Swann, a historian at the FDA, notes that the expansion of supplement use came at a time when people were taking charge of their own healthcare. Supplements give people a sense of autonomy over their health and fitness. They're something tangible that people can do to give themselves a promised edge, and if you're an athlete aiming for the top, you can't leave any possible edge untried. The idea that exercise creates extraordinary nutritional needs doesn't make a lot of sense when you consider that the body was made to move (being sedentary is what throws our nutritional needs off-kilter). Yet we are told again and again that our bodies need special nutritional coddling when they're active.

But more than anything else, supplements have become popular with athletes because the industry has bought its way into the culture of sport. Supplement makers bring money to the table, and they spread this money around—to organizations, coaches, trainers, teams, and individual athletes. Along the way, a lot of people get a chance to make a buck. It's not unusual to find trainers selling supplements to athletes, and gyms and sporting goods stores may get in on the act too, because the products can provide steady income. As Paul Bragg pointed out back in the 1940s, if you can convince athletes that they need your pills and powders, you can create a steady stream of income.

A recent report by the International Olympic Committee said that the reasons athletes cite for using supplements "are often based on unfounded beliefs rather than on any clear understanding of the issues at stake, and may reflect encouragement from individuals who are influential rather than being experts on this topic." In other words, athletes are being encouraged to take supplements by people who may themselves be under the influence of advertising or vested interests. While USADA is telling athletes in one ear that taking supplements puts them at risk of failing doping tests, the very companies that make these risky products are whispering to them in the other ear—here, have some sponsorship money and free products. Meanwhile, teams, leagues, and sports federations accept money from supplement companies, which in turn feed the cycle of FOMO by lending their legitimacy to these products. Although the people at USADA never say so, I can hear the frustration in their voices. Their message of caution is drowned out by marketing messages and sponsorship dollars. They're fighting a losing battle.

A 2017 study found that most athletes get their information about dietary supplements from coaches, trainers, friends, and family members.[30] Most coaches and trainers have little scientific training, and supplement companies blast them with promotional information about nutrition. These marketing messages usually fall on welcoming ears. It's appealing to think that something as simple as a pill or a shake might give us an extra something. With the marketing-based science giving the claims a sheen of truthiness, it all seems so plausible and potent. There's a reason hucksters have flourished for centuries—we are all too eager to believe.

9

Losing Your Zoom

With a scrawny distance runner's build and a mop of blond hair, Ryan Hall was once America's great hope for the marathon. He was not just talented but supremely driven. Hall grew up in Big Bear, California, with a father who was a marathoner and triathlete. As a middle schooler, Hall wanted to play football or basketball. "I hated running," he says. Then one day in eighth grade he was headed to a basketball game when he found himself fixated on a nearby lake. "It was like God was just giving me a desire to run around it," says Hall, whose religious faith plays a dominant role in his identity. "Something in that moment—the desire was awakened in me. I can only explain it as a God thing." He told his dad what he wanted to do, and the following Saturday, they set out around the lake, a 15-mile journey. "We were just plodding along. It was a really painful, long run." His choice of footwear—basketball shoes—didn't help. "I was all blistered up after that." It didn't matter. He was hooked.

"I loved the feeling of going all out," Hall says. "My dad wrote me weekly training programs and I was always pressing him to let me do more. I always wanted to do more." It's a common reaction among newbies—the thought that if a little bit of training gets you a small measure of success, a lot might get you much more. By his junior year of high school, Hall was running on the order of 80 miles per week, and during the summer of his senior year his weekly mileage reached into the 100-mile range, a number more typical of an elite marathoner than a high school runner. "Even when I was training for shorter events, I was always a big volume guy," he says, referring to the high volume of miles he put in throughout his career. He pushed himself hard, always, and for a while this approach paid off. As a high school track and cross-country runner, he won multiple state championships and set a California state record in the 1,600 meters.[1] He also finished third in the 2000 Foot Locker Cross Country Championships, the premier high school event. At Stanford University, he struggled with injuries, but he finished second in the 2003 NCAA cross-country championships and earned an NCAA outdoor track title in 2005 with his win in the 5,000 meters.

After college, Hall blazed his way into the record books. In 2007 he became the first American to break one hour in the half marathon. His time of 59:43 was fast enough to win the US Half Marathon Championships in Houston, where he crossed the finish line with a fist held triumphantly in the air. Most runners gradually work up to the marathon over a matter of years, but Hall already had his sights set on the distance. Later that same year he set another American record, this time for the fastest marathon debut, by running 2:08:24 (good for seventh place) at the London Marathon. Despite the record, Hall says it wasn't one of his best performances. "By the time the race rolled around I felt a little too far down the road in my training. I felt a little bit stale." That bit of fatigue was an inkling of things to come, but he brushed it off and kept pushing. His eyes were on the Kenyans and Ethiopians who dominated the marathon, and he was determined to keep going full throttle until he could run with them. "From the very beginning, I felt like this was my calling. I knew what my destiny was going to be," he says. His first (and only) marathon victory came in the 2008 Olympic marathon trials, and he went on to finish tenth at the Beijing Olympics in 2:12:33.

Hall's standout race came in 2011, when he ran the fastest marathon ever posted by an American, 2:04:58, finishing fourth at the legendary Boston Marathon.[2] For the first time in three decades, an American-born runner had become a legitimate contender in the marathon, an achievement that provoked comparisons to US marathon legends Bill Rodgers and Frank Shorter. All eyes were on him.

The autumn after that breakthrough run in Boston, Hall ran a 2:08:04 at the Chicago Marathon, good for fifth place and still one of the ten fastest marathons by an American. After that, things started to fall apart. He finished second in the 2012 Olympic marathon trials, but then dropped out of the London Olympic marathon before the halfway mark, and he withdrew from that year's New York City Marathon too, blaming fatigue.[3] The following year, he planned to run the Boston and New York City Marathons, but didn't start either one, owing to injuries. He didn't finish another marathon again until 2014, when he crossed the line at the Boston Marathon in twentieth place, about nine minutes slower than in 2011. The following year, Hall attempted a comeback at the Los Angeles Marathon, but after a blazing start in which he went out with the leaders, he abandoned the race near the halfway mark, once again vanquished by a nagging fatigue.

Hall found himself in a cycle he calls the "walk of shame." He would lace up his running shoes, head out for a run, and within 15 minutes his body would shut down, as if his battery had run out of juice. Unable to continue, he'd turn around and walk back. "I felt like I was allergic to running, like I was melting into the ground," he says. He had fallen into a pit of fatigue so deep he couldn't crawl out. "My body was sending signs. *I've given everything and there's nothing left to give.*"

By the time Hall retired from racing in January 2016 at age thirty-three, he felt he'd burned all his matches. He saw before him a choice. "I can keep struggling with this for another two, three, maybe four years and keep running worse and worse, or I can move on with my life." Framed like that, the decision to quit came as a relief.

———

What had happened to Hall? How had he gone from the great American marathon runner to something like *Groundhog Day*, stuck in that never-ending walk of shame? He may have felt run-down, but he wasn't sick. "I was getting my blood tested and nutrition panels and there was nothing pointing me to, if you fix this then you're going to feel so much better," he told me. Instead, he'd fallen into a cycle of diminishing performance most commonly referred to as "overtraining syndrome."

The current thinking holds that overtraining syndrome is what happens when the stress of training no longer provokes adaptations, and instead throws an athlete spiraling into a prolonged state of fatigue. Hall was training hard, but rather than getting stronger and faster, he was becoming more and more tired and couldn't rebound to his previous level of performance. Instead of adapting, his body cried "uncle!"

You can tell an athlete with overtraining syndrome because they "lose their zoom—they no longer have the magic that made them good," says Carl Foster, director of the Human Performance Laboratory at the University of Wisconsin–La Crosse and an author on the 2013 joint consensus statement on overtraining syndrome from the European College of Sport Science and the American College of Sports Medicine.[4] The overtrained athlete is more than just tired. "You let them rest for a few days or a week and they still haven't recovered whatever it is that makes them magic," Foster says. They can no longer perform, but there's nothing objectively amiss. "You send the person to a doctor who'll find that there's nothing wrong. They're not ill, nothing's broken, but they're just not right."

An unexplained drop in performance remains the hallmark symptom of overtraining syndrome. Yet even after decades of study, there's still no definitive criteria or test to identify it. Instead, it's a diagnosis made by ruling out other possibilities. In Hall's case, his performance plummeted for no obvious reason, and when anemia, hypothyroidism, and other possible illnesses or conditions were ruled out, overtraining remained as the most probable cause. Although his testosterone tested low, Hall told me that it has always been low and is probably just his natural state, since it hasn't budged in retirement, despite doing all the things, like gaining weight and lifting weights, that are supposed to raise it.

What makes overtraining so tricky is that its symptoms, which also include muscle soreness, extreme fatigue, sleep problems, and mood disturbances (most often depression and/or anger), are very similar to those normally produced by very hard training. Athletes might experience any or all of these things during a training camp, for instance, and ordinarily these are a sign that the training is having the intended effect. Blocks of hard training are meant to push an athlete into a state of overreaching, where the body is stressed beyond its current abilities in order to induce a training adaptation. When athletes are overreaching, they're pushing hard, and they feel it. During the period of intense training, they'll be too tired to perform at their best, but after some rest, they'll usually rebound.

The difference between the states of overreaching and overtraining can only be seen in retrospect, by noting how long they endure. So-called "functional overreaching" tips the athlete into a condition of fatigue, and this exhaustion may continue for a little while. But the key is that it doesn't last—after a rest period of several days or maybe up to a few weeks, the athlete's performance bounces back to even better than before. Runners get faster, strength athletes get stronger. With overtraining syndrome, however, the athlete's performance doesn't return to baseline, and the latest research suggests that it usually takes at least six months to get over it, Foster says. "That's *if* you're going to get over it. For some people, their careers are over."

Training alone can't explain the condition, because it can strike one athlete even as another on the same team, undergoing the exact same training regimen, doesn't succumb. What that says is that, despite its name, overtraining syndrome isn't a failure of training, it's a failure of recovery. For some reason, the athlete's ability to adapt has broken down and the body can no longer absorb new training. According to Foster, "We're pretty sure it's not training per se that causes it. It seems to be total life stress."

"There's a belief out there that you can't overtrain, you can only under-recover," says Shona Halson, the physiologist at the Australian Institute of Sport. "If you're sleeping well and you're eating well and you have minimal life stresses, then potentially you can take a larger and larger training load." Recovery is the limiting step. To get this point across, some researchers in the UK have proposed renaming overtraining syndrome "unexplained underperformance syndrome" or UUPS (pronounced like the plural of the word "up"). This change would take the emphasis off of training and encourage people to focus their attention on the factors obstructing recovery.

It turns out, the body's ability to recover can be hindered by a multitude of factors—insufficient sleep, not enough rest between hard efforts, poor nutrition, a bothersome cold virus, or, commonly, psychological stress. To the body, Halson says, stress is stress, whether it comes from a hard workout, a competition, a romantic breakup or, if you're a student-athlete, the anxiety of final exams.

———

So was Ryan Hall simply under-recovering? He is the first to admit that recovery wasn't his strong suit. He liked to push hard, and felt antsy when he rested. "I hated taking breaks," he says. "You're so accustomed to striving after things that to be able to relax is a big challenge."

Hall had begun his marathon training under the guidance of Terrence Mahon, known for coaching marathon greats Meb Keflezighi and Deena Kastor. "The primary job of a coach is to hold the athlete back," Hall says, but as an athlete, he admits that he didn't always comply. Mahon instructed Hall to take two weeks completely off following every marathon. "After I'd take these two-week breaks after the marathon I felt like I was completely starting from scratch." Following the London Marathon in 2007, Hall decided, "I'm not going to take a bunch of time

off. I'm just going to keep things rolling." He'd worked so hard
in the buildup to that race that he was afraid that by resting he
would lose his hard-won fitness. "By the time the summer came
along, I was feeling terrible and getting slower and slower."
With his upcoming Olympic trials marathon in jeopardy, he
was finally forced to take a week off to recoup, but it came at a
crucial training time.

Looking back at it after his retirement, Hall wishes he hadn't
skipped those prescribed rest weeks. If he had it to do over, he
says, "I would force myself to take two weeks completely off and
get fat and out of shape—that worked every single time I did it.
I see those breaks are what allowed me to run as fast as I did."

Following a major bout of fatigue that forced him to withdraw
from the 2010 Chicago Marathon, Hall split with Mahon to
coach himself, with guidance from God. "I was calling it faith-
based training," he says, noting that he once listed "God" on a
drug testing form that asked for his coach's name.

Hall didn't rely solely on supernatural power, however. In
advance of the 2011 Boston Marathon, he sought advice from
an unlikely source—a triathlete coach named Matt Dixon.
Among the endurance sport community, Dixon has picked up
a reputation as the "recovery coach." Dixon grew up outside of
London and arrived in the United States in 1992 on a swimming scholarship at the University of Cincinnati. After college, he became a professional triathlete and came of age as an
athlete at a time when "high volume, heavy, heavy work" was
the standard. "I was overtrained relative to my results. I took
the mind-set that this is a hard sport and I had to work really
hard," he says. He had some success, competing as a pro at the
Hawaii Ironman, for instance. Encouraged by a culture whose
barometer of training success was the number of hours put in,

he drove himself into the ground. "I ended up with some form of chronic fatigue before my professional career had ever had a chance to evolve anywhere near my potential. I was done." The experience led him to take a step back as he shifted into his next career, as a coach. "I thought, this is ridiculous. People would say oh yes, recovery is important, but it was pure lip service. It was train, train, train, train, train, then have an easy day. Then train, train, train again. I thought, there must be a more pragmatic way."

Dixon's solution was a radical reframing. "I decided to make recovery more than just lip service. I made it part of the program." He encouraged a new mind-set among his athletes. "It's not nutrition as a stand alone, or sleep as a stand alone—these are as much a part of the program as swim, bike, run. They must be taken as seriously as any training session."

Hall was introduced to Dixon by Chris Lieto, a professional triathlete whose breakthrough second-place finish at the 2009 Hawaii Ironman came after hiring Dixon, who cut his training hours by about a third and put more emphasis on nutrition and sleep.

"When Ryan came to me, he was in a state of disrepair and fatigue," Dixon says. "Just in the words he was using to describe his running, you could tell that emotionally, he was tired. There was no doubt that Ryan loves running and that Ryan loves to work hard—you could tell that straight away. But the candle was burning dull. His confidence wasn't there." Dixon reduced Hall's training load and admonished him to make sure his easy days were easy enough. "He had me do a lot of training with a heart rate monitor and he'd say, don't go over this heart rate," Hall says. "For someone like me who likes to push hard, it was a really effective way to hold back."

Dixon also saw that Hall was struggling to manage his weight. "It's very common in runners," Dixon says. "It wasn't necessarily an eating disorder, but a quest to get to a specific target weight." The problem was that the target Hall had chosen made him fragile, not healthy. "Matt told me, 'I've never met an endurance athlete who's eating enough food.' In running, there's such a value to being light and skinny and looking like the Kenyans. Matt was like, 'listen, you need to fuel your training.'" Dixon instructed Hall to become more diligent about getting enough fuel before a workout and replenishing his energy afterward. "Fueling is a massive performance-enhancing thing," Hall says. "Looking back, I ran my very worst races when I was my lightest." Dixon remembers a conversation he had with Hall before his magic 2:04 race at the Boston Marathon. "His emotional race weight was something like 134 [pounds], and he said, 'I'm a bit worried, because I'm at 137.' I told him, 'I'd like to see you at 138.'"

The few months that Hall consulted with Dixon leading up to his breakthrough Boston Marathon seemed to put him back on track after a year in which he'd suffered his first major disappointment—missing the 2010 Chicago Marathon because his body felt too broken down to run. "I changed his mind-set and philosophy and he did very very well," Dixon says. "But I think the question is, did he integrate these new habits for the rest of his career? And it doesn't seem like he did."

After working with Dixon, Hall turned to a highly sought-after Italian coach, Renato Canova. "He's coaching the fastest guys in the world," Hall said at the time.[5] He was intrigued. "What is he doing that's producing this many quality athletes?" It turned out that Canova's program was much harder than anything Hall had ever done before, and he soon felt himself "flirting with overtraining." Still, he looked at the Kenyans, like world champion marathoner Abel Kirui, who were flourishing under Canova and thought, "You have to be flirting with overtraining, that's just going to come with the territory." Hall doesn't blame Canova for his struggles. The two worked together from afar, not in person, and Hall says that this long-distance relationship surely limited Canova's ability to perceive how he was faring on the plan.

In hindsight, it might seem puzzling that Hall would abandon the recovery-focused approach that had resulted in his best-ever marathon to pursue a program that almost immediately put him back in the hole. But the thing to understand is that Hall wasn't satisfied with what he'd already done. He was never going to rest on his substantial laurels. He believed from the start of his career that he was destined to run with the world's best, and he wasn't going to stop pushing until he got there.

It's common, Foster says, for athletes on the road to developing overtraining syndrome to make things worse. "They're feeling tired and worn out, and their performance is starting to fall out. Almost every athlete and coach responds to failure with more effort, when what the athlete really needs at that point is a vacation." As driven as he was, it's no surprise that Hall fell into this pattern. "With Ryan, his greatest strength was his ability to grind things out," says Steve Magness, author of *The Science of Running* and coauthor of *Peak Performance*. "You see it a lot, not only in the best athletes, but in people who are really good at anything—they have this obsessiveness," Magness says. "You become obsessive at whatever the task, and the problem becomes not doing the task, the problem is shutting it off. Ryan fits that mold." In the end, Hall's greatest strength may have also become his ultimate downfall.

———

Hall never did get his zoom back, but some athletes with overtraining syndrome do. US national champion triathlete Jarrod

Shoemaker knew something was wrong when, during a winter training camp, he went to brush his teeth and put the toothpaste on his finger. "I was *that* tired," he says. It was December 2012, and Shoemaker, who had competed in the 2008 Olympics in Beijing, was in his first season working with a new coach. "It was the hardest I'd ever trained in December," Shoemaker says. His coach was trying to push him in ways he hadn't been challenged before, and that included training with other athletes instead of by himself. But rather than energizing him, training in a group setting seemed to sap Shoemaker's strength. "All of a sudden, it was a dog-eat-dog mentality," he says. "A lot of people have to win every workout, and being in that headspace, I almost used up all my mental energy training. I didn't have anything left to race."

Usually a good sleeper, Shoemaker found himself tossing and turning at night. "Sometimes I'd wake up feeling like I'd run a marathon while sleeping. My body felt really hot all the time, even at night." His breaking point came at the 2013 ITU World Triathlon in Kitzbühel, Austria. "I vividly remember standing there on the starting line, ready to go, and I was feeling like, I just don't care. I dove into the water, but I thought, I just don't care. I don't want to fight." He finished the race, then headed home to Florida to figure out what to do next.

Eventually, he connected with Neal Henderson, a former pro triathlete and coach to numerous Olympic endurance athletes. Shoemaker had already taken several months off, but Henderson had him wait until his mental and physical energy had returned before he resumed training. At that point, Henderson started him on a program of drastically reduced hours and had him focus on mechanics and form while he regained his energy. "He started me off on a super low burn for the first six months,"

Shoemaker says. Henderson was strict about when Shoemaker could push and when he should take it easy. It didn't happen overnight, but eventually Shoemaker came back. In 2014, a little more than a year after his crash and burn, he posted a personal best placing at the world championships.

Talking with Shoemaker, the difference between his experience with overtraining syndrome and Hall's seems striking. Where Hall responded to fatigue by adopting new workouts or exerting more effort, Shoemaker let go. That difference may have proven key. For athletes stuck in the kind of fatigue cycle that plagued Hall at the end of his career, there's only one solution—a long period of rest, says Foster, the human performance researcher at University of Wisconsin–La Crosse. "The athlete says, 'Well, if I rest a long time, I'm going to miss the game and I'm going to get out of shape.' Yes, you are, because you've got something that's made you where you're not any good anymore." Shoemaker accepted this reality and allowed himself to stop striving long enough to recover. Hall, on the other hand, never quite managed to feel at ease resting. He was a relentless striver to the very end.

———

Professional athletes like Hall and Shoemaker are naturally prone to overtraining syndrome because of the long hours they put in, but the condition is not confined to sport's elite. Henderson says that it may be even easier for age-group and masters athletes to fall into overtraining syndrome, because they have less time available for recovery and more things in their lives competing for their attention.

Kristina Kittelson is the former director of the Colorado Plateau Mountain Bike Trail Association and founder of a women's

cycling group in western Colorado. She's also a formidable athlete who can finesse her mountain bike through gullies of boulders and jagged rocks that make most mortals click out of their
cleats and hike-a-bike. She's the kind of riding partner who's
always suggesting we go a little farther or try a new loop.

But one autumn day when I met her for a mountain bike ride
in the mecca of Crested Butte, Kittelson seemed uncharacter-
istically subdued. She kept our easy pace as we pedaled through
the golden aspen trees, but she lacked her usual verve. Her
normal enthusiasm was replaced with a sense of resignation.
I started to understand what was happening when I asked her
how her summer had gone. "Not great," she said. She'd spent
the previous winter and spring training for a big mountain bike
race on her home trails. The year before, she'd entered the race
on a whim, and she'd won. She'd always ridden a lot, but had
never raced much. Her surprise win left her eager to see what
she could do, so she'd entered this year's race, and this time she
bumped herself up a category to a distance that was 10 miles
longer than the race she'd won the previous year.

It was time to get serious, so on the recommendation of
a friend she logged on to a coaching website and purchased a
training program designed for a 50-mile bike race. She didn't
get any real interaction with a coach, but she received a training plan that was specifically geared toward racers in her age
cohort. It seemed like a solid program, so she went with it.

The formal training plan marked a dramatic departure from
her usual routine. Before she'd started the plan, she said, "I'd
just go out and ride whatever pace and distance I wanted to. I
wasn't focusing on intervals, I was just riding." The plan didn't
increase her mileage. "It felt like I was riding less, because the
rides were shorter and emphasized intervals." The training was

more focused and higher intensity than what she'd done previously. Sticking to the plan felt stressful, not because the workouts were too hard, but because she had to adjust her life around them. She often rides with her husband, who has little interest in racing. "We'd go out for a ride and I'd need to do some certain thing, but he'd just want to ride. That just added to the stress."

Having committed to doing the race and told people about her intention, she felt compelled to do the training. "I was going to be the local person who does well in the race, so I put a lot of pressure on myself." Her first inkling that something was wrong came one day when she did a training ride on part of the course. Her route included some single-track riding and then a 10-mile road climb. "I did the single track and felt great and then I just died on the road. I didn't have any energy and I couldn't even finish the rest of the ride. At first, I thought maybe I'm just tired or it's because I'm having my period or something." But the feeling of fatigue persisted. "I wasn't sure if I was overtrained or undertrained, but I just knew that I wasn't feeling right." She felt moody, and she no longer enjoyed riding, nor was she making any progress in her performance. Her motivation to train was gone, and when she did ride, her legs felt heavy and she struggled to turn the pedals. She wasn't sleeping well, and that just sank her mood even lower. Meanwhile, her performance had plummeted.

Talking to Kittelson about her experience, I couldn't help wondering if perhaps the two ways of training that she'd tried—doing whatever she felt like on a given day and following a rigid training plan—were two opposite ends of a spectrum. Was it possible, I asked, that the training plan she'd adopted had pushed her over the line, not because of the specific workouts the program had prescribed, but because when she adopted it, she stopped making training decisions based on how she felt and instead allowed the plan to dictate what she would do? Could it be that she'd stopped listening to her body and thus failed to adjust to what it was telling her? Perhaps she'd have done better if she'd tweaked the training plan so that it didn't cause her so much life stress? Kittelson thought about it a little bit and conceded that it was possible, but she has no way to go back and try the same experiment again a different way.

One thing we know is that overtraining syndrome or UUPS isn't just a matter of training stress, but life stress too. I wondered if Kittelson's body was able to absorb more training when she was riding for fun, because at that time cycling was a source of joy and relaxation, whereas once she focused on racing it became work—a source of stress instead of recreation. I ran this theory by Kristen Dieffenbach, the sports science professor at West Virginia University who coaches numerous elite and age-group endurance athletes. She agreed that I was on to something. "It's a really fascinating paradox," she says. "All the research shows that exercise can be an excellent mitigator of stress, and it can help with mood and energy." But what can happen in a situation like Kittelson's is that what started out as a hobby and a source of relaxation and stress relief suddenly becomes a new stressor that actually increases the need for recovery, both physical and psychological. "I will often require athletes in this situation to go find some other physical activity that's joyful to replace it," Dieffenbach says. "You need to find that balance. It can be going for a walk or riding your bike easy with a family member, but you can't take any tracking devices with you and it can't have a training purpose. Until you can find that balance and joy, you've lost an essential piece of your recovery." Kittelson ended up taking a break from serious riding over that winter, and the break, along with a period without race goals, helped riding feel fun again. Her performance rebounded too.

———

The day we went mountain biking, there was something else that Kittelson mentioned in passing that caught my attention. Prior to that dispiriting training ride she'd had on the race course, she'd been fighting some kind of infection. It wasn't the sort of illness that knocked her out and confined her to bed, but she just didn't quite feel right. She rested a few days, but then pressed on with her training. I wondered, could the virus have made things worse?

David Nieman is pretty sure the answer is yes. Nieman is an exercise immunologist and director of the Human Performance Lab at Appalachian State University, and he's been looking at the link between exercise and the immune system for a number of years. The first thing to know, he says, is that prolonged, heavy exertion can suppress the immune system, and this "open window" of impaired immunity can last between 3 and 72 hours. During this period of susceptibility, any crud that you're exposed to has an opportunity to grab a foothold, and so the chance of an infection goes up. Factors like insufficient sleep, stress, weight loss, or poor nutrition can exacerbate the problem.

But here's the crucial part. If the athlete gets sick during this period, the body has a harder time clearing the virus, and if the athlete ignores the symptoms and keeps training, there's a chance that the body will be unable to totally clear the virus. "Then you enter a sort of subclinical, subperformance zone that's prolonged," Nieman says. Some researchers call it "postviral fatigue syndrome." A few of Nieman's colleagues in Australia studied a small group of athletes who'd developed fatigue so deep that they couldn't recover for at least two to three years. It turned out that 85 percent of these athletes had exercised heavily while they had some kind of niggling virus.[6] It's just preliminary evidence, but it's intriguing.

Nieman recalls a friend who trained an entire year for his first marathon. The night before the race, he got sick with a fever, but he ran the race anyway. "He entered that zone," Nieman says. "He couldn't do anything. His hands felt like he had arthritis. He'd try to sleep, but he'd wake up unrefreshed and with low energy. He was never the same athlete again."

At the moment, this postviral syndrome idea is just a hypothesis, one that needs more testing, Nieman says, but he believes that it points to an important lesson—"If you're sick, you just don't push. You could slip over the edge."

The thing about the edge is that once you cross it, it can be very difficult to come back. Perhaps the most important thing that researchers have learned about overtraining syndrome is that once you get it, there is no cure. The best you can do is rest and hope you recover. For this reason, most of the research efforts on overtraining syndrome are now focusing on prevention, and that means looking for ways to measure recovery and ensure that an athlete's training load doesn't outpace the body's ability to bounce back.

10

The Magic Metric

t's a fundamental issue facing athletes at every level—how much should I push and how much should I rest? How do I know whether my body is adapting and getting stronger, fitter, and faster in response to training, or whether I'm putting myself under? How can I determine if I'm training too little or too much? When is fatigue a sign that I'm doing the work I need to get better, and when is it a signal that I need more time to recover?

These are notoriously difficult questions to answer, because there are few universal rules that apply to every situation. "With apologies to Tolstoy, you could say that all recovered athletes are the same, but each unrecovered athlete is unrecovered in his or her own way," says William Sands, a sport physiologist at the United States Ski and Snowboard Association. Sands has studied overtraining for decades and ran the very first recovery center at the US Olympic Training Center in Colorado Springs. During the era of dot matrix printers, Sands coached gymnasts and he developed a computer program to track his athletes' training and recovery.[1] Each day, the gymnasts completed a questionnaire that collected twenty-one measures—everything from their weight to sleep times, psychological feeling prior to practice, health status, and the number and types of gymnastics moves they'd done that day. These measures were submitted on dot sheets that Sands scanned into the computer. He then applied artificial intelligence to alert him when one of his athletes had an anomaly that might suggest a problem.

During the twenty-five years that Sands spent monitoring gymnasts, he found that his athletes responded "characteristically, but idiosyncratically" to stress overload. Athletes' bodies can respond to stress in an almost infinite number of ways, yet an athlete's peculiar combination will be unique to them and when they face a new stressor, they'll tend to cope or respond to it in the same way, Sands says. Some athletes might have trouble sleeping when they're overstressed, others might lose or gain weight, or become moody or come down with a sore throat or cold. Whatever the response, he found it was usually characteristic for that particular athlete, but not necessarily universal.

As he says this, I realize that I have my own version of this phenomenon. It took many years, but I've come to realize that when I wake up in the morning with even a hint of a sore throat, it's a sign my body is stressed and I need a day off. If only I hadn't needed a decade of experience to recognize this sign!

What shows up as a sore throat in me might manifest as a particular kind of moodiness or a specific soreness or headache in another. Researchers have spent years looking for the magic metric that can quantify recovery and predict overtraining (or, better yet, help prevent it). No one has found the perfect number, at least not yet. "What makes studying the whole thing an absolute mess is that if you've got a group of ten people and you're looking at stress response in terms of heart rate and not everyone responds the same way, you won't see it," Sands says. "I'm hesitant to reduce it to one or two variables. There's too much individual variation."

Despite the challenges, companies across the globe are spending millions of dollars searching for the magic metric and convincing athletes they've found it. Irishman Brian Moore is a former middle-distance runner who runs Orreco, a startup company that looks to biomarkers in the blood and other tissues to track how the body is adapting (or maladapting) to training. Moore's company is built on a shining hope: that data can solve recovery's most vexing problem. It's just one of many companies that are betting that biometric data can provide individualized answers about how to manage training and recovery. With help from IBM's artificial intelligence program, Watson, Orreco is working on ways to use objective measures to help athletes and their coaches distinguish what they need to reach peak performance. "We're agnostic about where data comes from," Moore says. "Our product is insights." That last claim is a common talking point among tech companies, which routinely promise to deliver special knowledge that will change *everything*. The problem with technology is that it can lure us into mistaking the numbers we can collect with answers to the questions that matter. They may look the same, but they rarely are.

The reality is, blood tests don't automatically provide any new actionable information. "A blood test is not guaranteed to tell you anything at all," acknowledges Charles Pedlar, Orreco's chief scientific advisor. "It may not be measuring the right thing, or the problem you have might not turn up in a routine blood test."

One clear, actionable problem that's easily seen on a blood

test is iron deficiency. "That's one test that's very solid," Pedlar says. Low iron stores can cause fatigue, and endurance athletes are particularly susceptible to iron deficiency, especially menstruating women and anyone who trains at altitude. Furthermore, exercise like running that involves pounding may destroy some red blood cells in a process called exertional hemolysis. (Whether or not this takes out enough blood cells to create problems remains uncertain.) Fatigue can set in before an athlete has reached a clinical state of anemia, and a simple ferritin blood test can catch this.

Other tests are more nuanced, however, so you have to perform lots of blood tests to tell what you're looking at. For instance, Pedlar says, immune function may become suppressed over time as an athlete gets more and more fatigued, but tracking this with a blood test is tricky, because immune-related markers like white blood cell counts can change rapidly and for a variety of reasons. A change in cell counts may indicate a minor infection or it might indicate a natural suppression of white blood cells that can occur as you get fitter. A single measure is almost impossible to interpret, unless you're tracking the numbers regularly, over time. And without a larger context, the numbers don't offer much insight.

Despite these issues, numerous companies offer blood tests to any athlete willing to pay. Where Orreco works mostly with elite teams, Inside Tracker and Blueprint for Athletes both offer blood tests aimed at any athlete who wants, in the words of Inside Tracker, to "take a selfie from the inside." The message is that the tests can reveal crucial details about what's happening inside your body, information they imply could be otherwise invisible, yet crucial to your health.

It sounds very cutting-edge, but after talking with these companies, I wonder if they are going about this all backward—by taking existing technologies and looking for new markets for them, rather than examining a problem athletes face and looking for an optimal solution. It's the equivalent of a drug company that goes looking for the most lucrative, widespread problem it can treat with a drug it owns, instead of focusing on a disease and seeking the most effective treatments for it. It makes lots of sense for the company and its stockholders, but might not serve the patient so well.

Gil Blander, founder and chief scientific officer at Inside Tracker, all but acknowledged this to me. He says that initially he had hoped to develop tests that could help people with diseases like diabetes, but then he realized that "they are not the early adopter population." On the other hand, "the weekend warrior who will buy a $10,000 bike will pay a lot of money to shave a little time off his ride." He'd found his customers. Blander claims that hundreds of pro athletes use Inside Tracker, but says that the company's target market is mainly people who are athletic, but not necessarily pros—CrossFitters, Spartan Race athletes, runners, and triathletes, for example. He hopes to also find users among busy professionals and people who are health-conscious but not competitive athletes. Eventually, he would like to help people with serious health problems too. "It's the population who's sick that actually gave me the vision to start the company."

To develop Inside Tracker, Blander looked for blood markers that met three criteria. First, they should be biomarkers of health, not disease. Second, they should be factors that are less than optimal in at least 1 percent of the population. And finally, they should be things that can be altered with nutrition, exercise, supplementation, or lifestyle changes. The program offers tests on nineteen biomarkers, including vitamin D, glucose, liver enzymes, iron, blood cells, and hormones like testosterone and cortisol.

What distinguishes Inside Tracker from other tests is the way the company individualizes results and offers advice on how to improve them. According to Blander, most "normal" ranges you get on a blood test are "cookie cutter"—they're the same for everyone. "What we do is give everyone their own individualized zone based on age, and things like activity and the amount of alcohol you're drinking." When I asked how they did this, he explained that they used data from the Framingham Heart Study, a longitudinal study tracking a cohort of people from Framingham, Massachusetts, to study cardiovascular disease risks. The study has found what Blander calls "a nice correlation between low levels of glucose and low mortality," which forms the basis of Inside Tracker's optimal ranges for that marker at various ages.

All of the optimum ranges suggested by Inside Tracker come from peer-reviewed science, Blander says, or "from a big database of a quarter of a million people so we can say, 'Here's what your healthy peers look like.'" The company's recommendations are based on the population of your age, gender, and ethnicity. Blander did not reveal where this data set came from, just that it's "part of our proprietary intellectual property." I imagine that some people who buy the tests don't pay much attention to the fine print explaining that their data will be added to the database and become the property of Inside Tracker. Blander says that everyone who purchases one of their tests and contributes to the database may end up ultimately benefiting from it, but they'll still have to pony up to receive any benefits or insights gleaned from the data set they helped build.

Mike Wardian, a runner who holds the world record for a marathon run on a 200-meter indoor track, has been using Inside Tracker for about two years.[2] "If you do it once, it's hard to know much," he says. "The thing I geek out about is you can see trends if you do it regularly." A muscle strain or joint sprain may be obvious, but other problems are less so. "If something's hurting your liver or your blood glucose is out of whack, you won't necessarily see that. This is an opportunity to make some changes." Wardian is a vegetarian, and gives a lot of thought to what he's eating to ensure he's getting all the nutrients he needs. "I eat pretty well—lots of fruits and vegetables, and I'm always making sure to get enough protein and iron. But the ability to check on it is good peace of mind." The Inside Tracker results offer advice on how to improve blood markers that are outside of what they've tagged as your optimal range, and the company's software allows users to specify their particular likes and constraints. When it gives Wardian suggestions on how to get more protein, the foods it points to will be vegetarian sources. I asked Wardian what changes he's made in light of his results, and he said that he used to eat lots of blackberries, but switched to raspberries and blueberries at Inside Tracker's suggestion. He couldn't recall exactly why it told him to back off the blackberries, so I asked Blander. He didn't know the specifics of Wardian's case, but said the recommendation must have had something to do with fiber, which he said is "very good for metabolic-related markers like blood glucose and LDL or 'bad' cholesterol." Maybe. But according to the USDA nutrient database, blackberries have more than twice as much fiber as blueberries, and I'm skeptical that switching from one berry to another is going to change someone's life.[3]

Wardian doesn't give any one finding too much credence.

"You can't just switch from eating blackberries to raspberries and expect to get faster—you have to go to the track for that. But those little differences can be the difference—for me—between winning and losing, or for a lot of people between being able to finish a race or not." As for the extra zinc and magnesium he's getting by taking Inside Tracker's advice to switch from bran flakes to Total cereal, Wardian is pragmatic. "I don't know if I can tell the difference or not, but I want to make sure I'm at the optimal level." And that gets to the anxiety driving athletes to these tests. "You don't want to wait until you don't feel healthy," the founder of one blood test company told me. The fact that a whole industry has popped up to help healthy people find ways to feel anxious about their bodies seems like a statement about the weird times we're living in.

———

Blood tests are about more than just health. No one wants to miss a chance to gain an edge, no matter how small, and it's this fear of missing out that the tests offered by products like Inside Tracker and Blueprint for Athletes seem to exploit. The idea is that there's a better you out there, and these tests can help you optimize your life to become that perfect self. Blueprint for Athletes is a product of Quest Diagnostics, a Fortune 500 lab testing company, and the program gives them an opportunity to expand their market with tests that consumers can order (and pay for) themselves. (Quest also performs Inside Tracker's tests.)

I'd seen Blueprint advertised at several sporting events, and decided to give the test a try.[4] About a week after a friendly phlebotomist came to my house and drew three small tubes of blood, I received an email with my results and an invitation for a phone

call with Bunny Foxhoven, a registered dietitian and nutritionist and then senior clinical educator for Quest.

My results included values for forty-three different tests—
everything from red and white blood cell counts to glucose, tri-
glycerides, hemoglobin, cortisol, cholesterol, and vitamin D.
Two of them, estimated glomerular filtration rate (eGFR) and
creatinine, were marked in red. My eGFR was low and my creatinine was high. Both of these measures relate to kidney function, and I won't lie—that scared me.

Foxhoven told me that the eGFR number could mean that my
kidneys are out of whack and the high creatinine level could also
signal a kidney problem, or it could mean I'm eating more protein than my kidneys can easily process. She asked a bunch of
questions, almost all of them focusing on what I eat. What was
a typical breakfast like? How much protein and salt do I consume? I explained that I eat protein throughout the day, usually
eggs and nuts as well as meat four or five times a week. I admitted that I like salty foods, and I often add salt at the table.

She instructed me to cut back on salt and talk to my doctor.
She also said it was probably a good idea to get tested again in
three months. "You do want to get on it before it's at the point
where you can't repair it," she said of this kidney problem that
may or may not exist. Even more scary? I might not experience
any symptoms until my kidneys lost up to 90 percent of their
function, she said.

I contacted my doctor, who told me not to panic. My numbers
were only slightly off, and probably not a cause for concern. So-
called incidental findings like these are common when you go
looking for anomalies in healthy people, said Michael Joyner,
a physician and human physiology researcher at the Mayo
Clinic. Both of my measurements that were outside the normal

range had to do with kidney function, and the two were related. My creatinine is high, Joyner told me, because I've got some muscles built up. It's possible I was a little dehydrated too, but I also work out regularly and exercise raises creatinine levels because it creates a little bit of muscle damage, which releases creatinine into the blood. His explanation was the obvious one, he says, but if I'd brought the results to a doctor who was less aware of the problems that happen when you give medical tests to healthy people without symptoms (the "worried well"), I might have been sent onto a spiral of escalating testing that would make me anxious and suck up a bunch of my time and money without improving my health one iota.

Reading the fine print that accompanied my results, I discovered that even Quest cautioned against reading too much into the high creatinine levels. Elite athletes tend to have higher creatinine levels than the nonathletic population, my results packet said. "This is likely due to the higher muscle mass in athletes, as muscle mass seems to be the best predictor of creatinine production." I'm no longer an elite athlete, but I have a muscular body type and I'd gone on a trail run a few hours before my blood draw. The Blueprint results packet advised that athletes should compare their measures to their own baselines, not the reference range, which is based on the population at large.

My slightly high creatinine level meant it was almost a given that my eGFR rate would be off too, since it's calculated based on creatinine levels. "Some research has also shown that both endurance athletes and team sports athletes may have lower eGFR," the Blueprint results explained. And just like for creatinine, it advised me to compare my results to my baseline, rather than the reference range. In other words, my slightly low eGFR was probably not a reason to panic, but I should get more tests to be sure.

I asked Foxhoven to show me how my results could improve my performance and recovery. Should I just take it as good news that they'd found nothing wrong? Because, honestly, I didn't feel like I'd learned anything meaningful. "Okay, good points," she said. If I was overtraining my muscles and joints, the tests might show it in the levels of stress hormones like cortisol or C-reactive protein, a marker related to inflammation. What puzzles me is that when I was an elite skier and suffering from a bout of what I can recognize in retrospect was overtraining, my doctor sent me for some blood tests, which showed nothing wrong. If blood tests couldn't help me when I was in the throes of overtraining, how were they going to help me when I was feeling good?

The advice that I got based on my results—to watch my salt intake and focus on hydration—seemed a lot like a modern version of Ovaltine's Orphan Annie Secret Society decoder ring that Ralphie anxiously mails away for in the movie *A Christmas Story*. When the decoder ring arrives, Ralphie uses it to decipher the secret message: "BE SURE TO DRINK YOUR OVALTINE." It seems silly to go to the trouble of giving athletes a test whose results are just going to tell them to keep hydrated and eat their vegetables. When I told Joyner that the Blueprint educator had told me to focus on staying hydrated, he laughed, calling it "the world's best piece of generic advice to athletes."

I called Richard Schwabacher, executive director of sports and diagnostic solutions at Quest Diagnostics, to find out what I'd missed. How would these results boost my performance or health? "The question you're asking is a hard one to answer," he said. "It really depends on the results." I'd probably feel differently about my results if they'd shown some vitamin deficiency I could correct by changing my diet or taking supplements, he said. "The good news is you're healthy. The bad news is that sort of takes away some of the value of the test."

Some athletes who get normal results take comfort in knowing that the reason they feel healthy is that (what do you know!) they really are healthy. He told me about an athlete who received her test results right before a big event. All the numbers were in the normal range, and that gave her peace of mind and confidence that her body was in optimal shape going into the race.

There's nothing wrong with people seeking data on themselves. And the products that companies like Inside Tracker and Blueprint for Athletes provide are legitimate medical tests that are carefully done. But, as my results illustrate, when you give screening tests to healthy people, you may end up with more questions than answers.[5] Sure, test your blood for iron if you have symptoms of low iron, but if you test your blood without a clear purpose, you're essentially just fishing around to see if anything seems out of line. Measure enough things in the blood, and chances are, you'll find *something* that seems just a little off. Normal reference ranges use a 95 percent confidence interval, and that means that five percent of normal, healthy people will fall outside of that range. Calling these values abnormal is essentially a false positive, and because some of the companies that market tests to athletes use an even narrower reference range, that means they're going to generate even more false positives. (They're also, of course, incentivizing people to come back for retesting. *Ka-ching!*)

Tests like these highlight an important difference between tech industry "innovators" and medical experts, says Vinay Prasad, a hematologist-oncologist at the Oregon Health & Science University. The tech nerds assume that more information is always better, but Prasad says that people like him who practice medicine know that information isn't the same as knowledge. "Just because you have a number doesn't mean it's helpful," he says. These blood tests are a perfect example of the McNamara fallacy at work: "Not everything that counts can be measured, and not everything that can be measured counts."

Marketing these tests to athletes so that they can try to "improve" their biomarkers is the sporting equivalent of teaching to the test. Rather than teaching athletes to read their bodies and understand when they're tired and need rest, these tests draw attention to numbers that may or may not be relevant.

———

We are in the midst of a data explosion. In this era of smartphones, personal trackers, and the quantified self, athletes willing to shell out some money can collect an astounding amount of data on their body and their training. Tech companies are standing by with gadgets and apps that can measure everything from the number of steps you take, miles run, elevation climbed and descended, and heart rates, to power output, workload, sleep, weight, and even some nutritional factors.

Perhaps the most basic metric is morning heart rate. With nothing more than a watch and a finger on your pulse, you can measure how fast your heart is beating. The fitter you get, the lower your resting heart rate becomes, and athletes who do this every morning will quickly figure out their normal range. If you wake up with an elevated heart rate, that's a sign of fatigue. It's a useful measure, but in practice, it's a pretty crude tool for measuring recovery, because it only offers insight into one piece of the puzzle—how hard your parasympathetic nervous system is working. The parasympathetic system promotes relaxation and controls subconscious activities like breathing and digestion.

Heart rate variability, or HRV, is a more sophisticated metric that measures the time between your heart beats when you're at rest. "Most people assume that when you're at rest your heart should beat with a metronomic regularity, but that's not actually true," says Simon Wegerif, director of HRV Fit Ltd., the company behind myithlete.com. In fact, your body is a complex system always seeking the most energy-efficient equilibrium and as a result there's a lot of variation on the beat-to-beat timing of your heart. "This isn't random variation and it doesn't mean your body isn't working well," Wegerif says. Instead, it's your heart working efficiently and usually this means that when you breathe in, your heart rate gets faster and when you breathe out, it gets slower. "This isn't a small variation. In very fit athletes it can be almost a two-to-one ratio. By measuring this variation and quantifying it and baselining it for an individual, we can actually tell how much stress their body is under." High variability indicates that your heart is rested and working efficiently, while lower variability is a sign of stress. It's a holistic measure of stress that can detect stress whether it's due to training, emotional stress, or a lack of sleep.

The myithlete program uses an external sensor to measure morning HRV, and integrates it into a training program that makes recommendations based on your HRV fluctuations. But some other heart rate monitors will also measure HRV, and there are now numerous smartphone apps available that use the phone's camera, placed over a fingertip, to measure HRV. Most of these apps also offer some kind of tracking features.

Polar and Garmin both make sports watches that track activity with data like distance, time, and intensity, and the information is fed into their proprietary algorithms to produce recovery measures. After a workout, your watch tells you how much rest you need before the next one, and gives you a recovery score or ranking. I've spent close to a year testing out products from both companies. (Yes, I'm the dork out there running with sports watches on both wrists.) What I've found is that they rarely tell me anything I didn't already know. When I finish a workout, I don't need to look at my watch to know if it was hard or easy or how tired I'm feeling. If I've done something really hard today, I know that tomorrow I'll want to take it a little easier, and that's pretty much what these trackers tell me too.

Neither measure has struck me as more accurate than the other, and for the most part they're reasonably correct. Except when they're wildly off. After a nearly four-hour cross-country ski that was about an hour longer than any training session I'd done yet that ski season, the Polar told me my training effect was "mild" and prescribed only about a day's rest. Sure, my pace and intensity was low, but I was totally spent for several days afterward. And the Garmin recently rated one of my easy runs as threshold training. Why? Without understanding the proprietary algorithms that calculate the ratings, it's impossible to know. What I've noticed is that I find these scores most interesting when they seem wrong, but usually when this happens it's not because they're pointing me to something I've neglected to recognize, but that we're interpreting the same workout differently.

A measure that I've found far more useful (at least for endurance training) is the "Training Stress Score" calculated by Training Peaks, an online and app-based program for synching and analyzing training data collected from sports watches and apps.[6] The TSS is modeled after the TRIMP or "training impulse" score developed by E. W. Banister and T. W. Calvert in the 1980s.[7] Both TSS and TRIMP take into account the duration as well as the intensity of a training session to produce a score that can help track your training and give a picture of your progress and potential for fatigue. It's not perfect, but it's slightly more transparent than the proprietary algorithms. I like that it works regardless of the activity I'm doing, and I can use it to track the additive effects of my various skiing, running, and cycling workouts. Eyeballing my TSS score (which Training Peaks helpfully graphs) is an easy way to visualize how my workload is changing day-to-day or week-to-week, and it's easy for me to spot worrisome spikes.

The sports that I do—running, cycling, cross-country skiing— are generally measured in hours or kilometers, but some activities have sport-specific measures that can help monitor fatigue and recovery. In baseball, for instance, pitch counts have become a popular way to track training load for pitchers, whose shoulders and elbows are notoriously susceptible to career-ending injuries. Research has shown that young players who throw a large number of pitches are especially vulnerable, and too many pitches and not enough rest seem to put any player at risk.

There's little doubt that insufficient recovery contributes to injuries, including things like tendonitis, muscle strains, and stress fractures. But the underlying causes are often more complicated than that, says Mick Drew, a sports scientist at the Australian Institute of Sport. "The more you train, the more you get injured, but that's probably just a measure of exposure," he says. "We also know that if people train consistently, they're protected from injury." The trick is figuring out the optimal balance.

The studies on training load and injury risk are imperfect, Drew says, but they all show pretty much the same result—that the greatest risks lie at very low and very high training loads. "A training spike on a very low chronic load is a perfect storm for injury," he says, but "if you have a high training load we know you're protected—to a point. A super high load? Maybe not." The key, he says, is consistency.

Drew argues that what have traditionally been called "overuse injuries" could be more accurately deemed "training load errors," where training load is the training stress placed on the athlete's body.[8] Someone gets sick or has a minor injury, and they miss out on some of their training. Instead of taking care to ramp back up to the amount of training they were doing when this setback happened, they push their training load to where it was before the break (or even higher, because they're worried that they've fallen behind), and the cycle continues, he says. Athletes who fall into this boom-bust cycle tend to experience multiple illnesses and injuries. "What we're saying is, be consistent with your loads. The importance isn't the metric you use to measure them, it's what the metric is telling us." The best way to protect yourself from injury is by being consistent, only doing what you're prepared to do, and taking it slow when coming back from injury. High training loads aren't inherently risky as long as you reach them in a gradual manner. But when loads spike, so does your risk of injury and illness.[9]

Carefully monitoring training load might seem like an obvious solution to this problem, but the fix has proven more complicated than that, because it's not the external load—the distance covered or number of sprints, for instance—that matters most. It's the impression that the athletes have of the load, says sports physician Jan Ekstrand, who leads an ongoing injury study looking at soccer players in elite clubs in the UEFA Champions League and beyond. The exact same task can feel very different to the same athlete, he says, depending on how much sleep they've had, whether they have muscle soreness or they're feeling fatigued from a previous workout, or are stressed by events in their personal life.

Once a year Ekstrand's research group gets together with team doctors to debrief. As part of this process, Ekstrand's group asked them, What are the most important factors for injuries and injury prevention? The four reasons they came up with surprised Ekstrand. The first factor was the leadership of managers and coaches—How well is the team run? The second was the load on the players—how many matches they play and how much they train. Next was internal communication—Do players have someone to talk with? Do messages from leaders easily reach the other people in the organization? Finally, the doctors pointed to the overall well-being of the players—what's happening in their private lives, and whether they have friends and good relationships.

These answers came as an "absolute surprise" to Ekstrand. "Normally we think of injury prevention as doing some strength exercises," he says, but the factors the team doctors listed as stressors could be relevant to any workplace. When people are unhappy in their jobs or their personal lives, work performance drops. "We need to broaden our visions," Ekstrand says. "We cannot just look at doing some exercises, we must look at the situation as a whole."

As an example, he points to the way that different UEFA Champions League teams care for their players who come from countries outside of Europe. "Some clubs are very good—they arrange housing for the family, school for the children, language teaching for the family and friends for the wife and so on," Ekstrand says. The family feels supported as they transition into a new culture. But other clubs don't provide this level of help. "They say, you earn so much money you should be happy playing for this club. Then you have a family where the wife wants to go home, because she doesn't understand the language and has no friends." The player in this latter situation might have stellar support on the field, but then may come home to a stressful and unhappy situation. Emotional strain at home can hinder recovery and take a toll on performance on the field.

In light of these factors, Ekstrand's group has started analyzing the training load on players using a measure called RPE, or rating of perceived exertion. RPE is essentially a number used to answer the question, how hard does this bout of exercise feel?

———

For a long time, overtraining was mainly studied by physiologists who used the tools, perspectives, and mind-sets of physiology to examine the problem. "The viewpoint that psychology could play a role in overtraining was just scoffed at," says psychologist Jack Raglin, who began studying overtraining as a graduate student in the 1980s. That is, until Raglin's mentor, William Morgan, came along. In 1987, Morgan published ten years' worth of data tracking mood in competitive swimmers.[10] Morgan and his colleagues at the University of Wisconsin–Madison sport psychology lab recorded the mood states of swimmers over the course of ten seasons and found that mood disturbances—things like tension, depression, and anger—rose in a dose-dependent manner as the swimmers trained more and then returned to baseline levels when their training dropped. The mood questionnaire that Morgan's team had developed elegantly correlated to states of recovery, almost in lockstep. As recovery went down, so did athletes' moods, and when their recovery improved, so did their state of mind.

"That paper really opened people's eyes," says Raglin, who took part in some of the work and is listed as an author. Physiologists had been looking at things like testosterone and cortisol ratios, lactate and other physiological markers that "weren't really telling us anything," Raglin says. Mood, on the other hand, appears to be almost an index measure of all the inputs that go into recovery.

Raglin developed a short mood scale questionnaire for soccer players to use in the morning. "It worked," he says, "but it took us several months just to implement the system where the players would do it and understand why it was useful." Raglin soon found that athletes were more resistant to filling out a questionnaire than to donating blood or taking a physiological measure like heart rate. "The whole notion of recovery is not well appreciated or understood by many in the sporting world," Raglin says, and it can take work to get athletes to buy into the idea that tracking psychological measures can help their physical performance.

Sport psychologist Göran Kenttä has worked with several Swedish national and Olympic teams. He uses Raglin and Morgan's POMS, or profile of mood states questionnaire, to track his athletes' response to training. Kenttä recalls one member of the Swedish kayaking team whose POMS assessment suggested that she was fatigued and not responding well to her training going into a heavy training camp leading up to the world championships. Kenttä sat down with the athlete and discussed her POMS scores. "I could tell from the assessment that we needed to drop her training load," he says. But the kay-aker resisted, worried that if she rested more she'd lose her edge. She made Kenttä promise that her performance would improve if she rested instead of training. Kenttä assured her it would. He was *that* confident in the POMS test. While the rest of the team members spent the camp pushing themselves to the brink, she backed off her training and focused on recovery. The result? She came back from the world championships with a bronze medal.

—

When Boston-based tech entrepreneur Jeff Hunt started doing triathlons, he was shocked by the culture's focus on logging training hours and what he saw as a dismissal of recovery. "They were so worried about getting their time in—it was like an obsession. I wondered, why are you exercising?" Were they aim-ing to get an ideal training response, or just going out to rack up numbers in their training diaries? Seeing a need for a way to prioritize recovery, Hunt joined with a Nordic skier friend to found Restwise, a program and app for tracking physiological and psychological aspects of recovery.

The goal was to have a quick, simple, noninvasive read on your current state of recovery, one that takes into account not just training but other factors that can aid or impair recovery, Hunt says. The program collects information on about a dozen measures like sleep, mood, the quality and intensity of yes-terday's performance, wellness, muscle soreness, hydration, and energy level. The enterprise started out with a gizmo that measured oxygen saturation in the blood, but Hunt says they quickly realized that "it tells you almost nothing." They've like-wise played with HRV, but as Hunt points out, there are "a lot of weird anomalies between what HRV tells you and what common sense tells you. I've run three marathons, and all three times HRV told me to go out and train normally the next day." He's found that things like the quality of yesterday's workout, mood, sleep, and energy levels are usually most meaningful. The program takes the inputs that users submit and uses a proprietary algorithm to create a recovery score.

Restwise users have included endurance athletes like triathlete Jesse Thomas, multisport champ Mike Klosser, cyclist Rebecca Rusch, numerous CrossFit athletes, the US Field Hockey team, and triathlon coach Matt Dixon, who used the program with Ryan Hall leading into Hall's spectacular Boston Marathon in 2011.

"The real value of Restwise is as an educational tool," Hunt says. "It teaches athletes how to pay attention." And not just pay attention, but pay attention to the right things. "People get accustomed to checking in with themselves and paying attention to how they're feeling. They get to where they can predict what their score will be. And when that happens, we've totally accomplished our role—to help people avoid overtraining."

———

We tend to think of objective measures as more reliable or scientific than subjective ones. If we want to know how fast a car is going, the speedometer provides a more reliable indicator than the driver's subjective feeling of the vehicle's speed. But recovery is a far more complex phenomenon than speed, because it isn't a measure of something the body is doing at this very moment, but a prediction of how prepared it is to perform at a given time.

We love to think that hard numbers provide the best data,

but it turns out that subjective measures trump objective ones when it comes to measuring recovery. In 2015, Australian sport scientist Anna Saw and her colleagues gathered up the published studies on metrics used to quantify training load and response—everything from hormone levels to inflammation markers, blood cell counts, immune system markers, and heart rates. When they analyzed all the results, they found that subjective, self-reported measures trumped objective ones.[11] For many of the objective results, it was very difficult to determine whether differences from one day or week to another were the result of meaningful changes in recovery or training response, or just normal variation in that particular variable. "The extent of this finding was surprising," Saw told me. "I didn't expect objective measures to be that inconsistent or subjective measures to be so reliable."

———

Kristen Dieffenbach, the sports scientist at West Virginia University who coaches elite and age-group endurance athletes, likes to use a budget analogy with her athletes. The amount of physical and emotional energy that an athlete has represents her or his currency. "You start with an assessment—how much are you spending? How much do you have available to spend and how much are you already committed to spending elsewhere?" Athletes need to be realistic about how much currency they have. Spending it all on training is like blowing your monthly finance budget on shoes, she says. Suddenly you're unequipped to meet the other demands in your life.

A few years ago, Dieffenbach collaborated with Göran Kenttä on a study in which they interviewed a series of elite athletes about their experiences coming to terms with their rest and recovery needs. Almost every one of them said that they'd started out their careers taking recovery for granted. The importance of rest was something they had to learn, and it wasn't enough to have someone counsel them on the issue, they needed to truly believe it. "So many of them looked back and said, 'If I had been better at my recovery when I was younger, I might have been a lot faster,'" Dieffenbach says.

But the culture of "go hard or go home" and the advent of social media apps like Strava and MapMyRun that allow people to view data on other athletes' workouts means that "everybody is constantly comparing themselves," Dieffenbach says. "It becomes a competition for training, but in those apps you don't log everything else you did in the day. You don't know what anybody else's budget is or what genetic lottery she won or trust fund she was born with. Maybe she needs more sleep than you do. Those comparisons won't tell you that."

Too much focus on numbers can be counterproductive. "Very few people take into account the psychological ramifications of all this science-y data," says Steve Magness, the running coach and coauthor of *Peak Performance*. "The more you measure, the more you create athletes who are fragile. I've seen it at the largest stage with world-class athletes. They become dependent. If the fancy HRV or omega wave tells them they aren't ready, then they believe they are not ready."

What makes tracking and data analysis so appealing is also what makes it dangerous—it conveys a sense of certainty that the science cannot yet deliver. The assumption underlying the use of data to make decisions about when to train and when to rest is that we understand the complex ways that the body processes stress and recovery and how various workouts affect an individual. In reality, our current understanding of these processes is still pretty rudimentary. "First, we don't even have a great model on how adaptation actually works," says Magness. "Second, we can't even agree how to quantify the stress of a workout."

When he sees an athlete obsessing with tracking numbers, triathlete coach Matt Dixon worries. Although he encourages his athletes to use tracking data as feedback tools, he hates to see anyone rely too heavily on them. "The engineers and measuring fanatics will kick and scream, but performance is not a simplistic calculation. Disregard your internal clock and perceived effort, and you will *never* truly evolve as an athlete," he writes in his book, *The Well-Built Triathlete*.[12]

"Whenever an age-grouper asks, how many hours am I going to have to train? My answer is, I don't know," Dixon says. Until he gets to know his athletes—what their backgrounds are, what kinds of stresses they have in their lives, how they respond to different types of training—he can't know how much training or how much recovery time they'll need. Dixon tells his athletes that if they want to be successful, they have to learn to be aware of how they're responding to their training and develop the good habits required for recovery—adequate sleep, good nutrition, and proper fueling of their workouts.

One of the most important things an athlete needs is confidence to listen to their bodies and trust in their training program. There's no single path to success, but this is a hard lesson for athletes who are constantly comparing themselves to their peers, a habit that can become even more compulsive these days, when athletes share their workouts on Strava and other social media. Dixon recalls a conversation he had with one of his athletes, Tim Reed, who won the Ironman 70.3 world championships in 2016. Reed wanted to talk about how other champions had trained. "This is how Craig Alexander did it, and this is how Cameron Brown did it," Dixon says. "And I said, Tim, if you want to be world champion, you have to do it the Tim Reed way. You can look over the fence, but ultimately a great champion will do what's right for them."

The way that athletes respond to training isn't dictated by physiology alone. How someone absorbs training is also influenced by the other things going on in their lives. As an example of this, Dixon points to two of his triathletes who were age-group world champions at Ironman in Hawaii. "These guys were in the same age group, just a couple of years apart," he says. One athlete, Rich Viola, was on a training recipe of about 22 to 25 hours per week. He had the luxury of time and his job, while very high powered, wasn't too stressful. The other athlete, Sami Inkinen, was the CEO of a startup and only had a training capacity of about 12 hours a week. Inkinen became an age-group world champion on almost 50 percent of the training load of Viola.[13] "It's not that 12 hours is the appropriate training load for everybody. It's that if I had Inkinen putting in 16 or 18 hours per week, he probably would have failed, because he was running a startup tech company that was very challenging and he only had so much time," Dixon says. "It's an optimization challenge."

A native of Finland, Inkinen moved to Palo Alto in 2003 and started doing triathlons in 2004. He calls himself an "incurable data geek," and says he loved "the excitement of figuring things out" regarding his training and performance. In 2005, Inkinen cofounded the real estate website Trulia, which left him few hours to train. "So I started to ask, how do I maximize physiological gains with the minimum amount of time?" Here, Dixon provided helpful guidance, but Inkinen also looked to scientific literature for insights. "I realized that a lot of the cardiovascular markers that correlate with endurance performance can be developed very effectively with very high intensity training, which requires less time," he says.

As a numbers guy, Inkinen wasn't just going to read a paper, he wanted to know how *his* body was responding. "I need to get objective data that I'm super-compensating and getting better every week," he says. After years of tracking all kinds of information about himself, he's concluded that "the best algorithm for taking in all those data points is our own brain. The morning mood and feeling is probably the most accurate predictor of recovery—it factors in a lot of things: injuries, hormonal status, hydration, nutrition." In isolation, single measures, like heart rate, HRV, or body weight, only represent a single dimension of recovery, he says. Blood markers don't impress him. "A single biomarker at any given point is meaningless. If you feel like crap, that tells you something. We're a long way from having a system of biomarkers that can surpass the human brain."

———

Camille Herron is a superhero. Really—she set the Guinness world record for the fastest marathon run by a woman in a superhero costume by winning the 2012 Route 66 Marathon in Tulsa dressed as Spider-Man. (Her record time, 2:48:51, was less than three minutes off the 2012 Olympic trials standard, which she'd already met.) Herron is a very fast runner, but her superpowers are her supernatural talent for recovery and finely tuned ability to listen to her body.

Trained as a scientist—her master's thesis examined exercise and bone recovery—Herron grew up in Oklahoma and played basketball as a kid. Her father and grandfather played basketball at Oklahoma State, and that's what she aspired to as well. Then in high school her coach noticed that Herron could run almost nonstop. "It looks like you might be a distance runner," he told her. She'd found her niche. She went on to become a three-time high school state champion and the University of Tulsa recruited her to run on its team. Her college career was punctuated by injuries, but during college she met her husband, a professional runner who was training for the 2004 Olympic Trials marathon. With her college career over, she started running longer too, and pretty soon she was doing marathons along with him.

In 2008 she qualified for the Olympic marathon trials and in 2010 she won the Dallas Marathon. She'd been invited to run another marathon five weeks later, and, noticing that she seemed to be recovering really fast, she thought, "Why not? Let's just see what happens." She ended up doing four marathons in three months, all of them in times under the Olympic trials qualification standard. In 2011, she was the top American in the US Pan American Games Marathon and finished as the third American in the New York City Marathon two weeks later.

As time went on, she was doing more back-to-back races and she started to think about running even longer. She made her ultramarathon debut in 2013, and after an injury-prone first year, she was soon winning races. She holds the world record for 50 miles and was the world champion at both 50 km and 100 km in 2015. In 2017, she became the first American in twenty years to win the storied Comrades Marathon, an ultramarathon between the South African cities of Durban and Pietermaritzburg, a distance of about 56 miles.

Tall (5 foot 9) with long, lanky legs, Herron has a free-spirited running style that is unmistakable, even from a distance. She runs with her below-shoulder-length blond hair flowing untamed as she glides down the road or trail. (About fifteen years ago, she decided to stop fussing with her hair. No more ponytails and hair clips.) If there's one thing that's apparent from watching her, it's that she runs by feel.

"What I've found works for me, is on a regular training basis to push myself to the point that I feel tired," Herron says. She usually runs twice a day and can do six or seven days of doubles like that before fatigue sets in. When that happens, she takes a rest. "Rather than planning my day off, I'm intuitive with what my body's feeling," Herron says. She races until she feels herself becoming tired, and then takes a few weeks or even a month off from racing, while easing off on the training.

"I just need to go with the flow of my body. When it's tired, I back off and then go at it again," she says. "What I've learned is that rather than planning everything on paper, and thinking that you're making the right choice, you have to keep a focus on yourself and what you're feeling—that's more important than a plan you wrote down in advance."

Despite her unusually fast recovery, Herron isn't invincible. In 2017, she won New Zealand's 102 km Tarawera Ultramarathon in course-record time, and after her Comrades win she aimed for the Western States 100, one of the most competitive ultras in the United States. She was a prerace favorite to win but ended up dropping out after about 15 miles. Her hamstrings had flared up, and she was slipping on the trail's many snowy sections that remained after the previous winter's epic snowfall. At the aid station where she quit the race, "We cried, laughed, shared a beer, and took pictures. The sunrise with the snow was like nothing I've ever seen," she told *Trail Runner* magazine, adding that she was at peace with her decision to drop out.[14] "It wasn't meant to be, and I will definitely try again." She thanked her supporters for their help, but did not apologize for the disappointment. She knows who's boss. Her body decides what she can do—not outside pressures.

Her race at Western States ended in disappointment, but calling it quits surely helped facilitate her next big accomplishment—a first-place finish and new world record for 100 miles (the fastest ever run by a woman or man on a certified course), set at the Tunnel Hill 100 in Vienna, Illinois. While Watson and all its brethren search for the perfect algorithm for balancing exertion and rest, Herron has perfected nature's own metric—a keen sense of how you feel.

11

Hurts So Good

'm writing this at a desk cluttered with a pile of recovery products I've collected over the months—hydrogen effervescent magnesium pills, a muscle roller, a sand belt–like vibrating muscle massager, a certified kosher gluten-free recovery drink, mechanized footbeds that push on your arches in 30-second cycles, and an oxygen inhaler, to name just a few. After exploring a seemingly endless array of recovery aids, I've come to think of them as existing on a sort of evidence continuum. At one end you've got sleep—the most potent recovery tool ever discovered (and one that money can't buy). At the other end lies a pile of faddish products like hydrogen water and oxygen inhalers, which an ounce of common sense can tell you are mostly useless. (It's not the amount of oxygen you're breathing in that usually limits your performance and recovery; it's your body's ability to shuttle it to your muscles, and inhaling shots of oxygen won't help that.) Most things, however, lie somewhere in the vast middle— promising but unproven. Again and again as I looked into the science, I found that the research on most recovery modalities is thin and incomplete. Maybe there's something to them, but it's hard to say for sure.

When these recovery modalities do work, their benefits are usually measured in single digits, and there's an "inconvenient truth" here, says recovery expert Shona Halson.[1] The small performance-enhancing effects of popular aids do not seem to add up to larger effects. Instead, there seems to be a plateau effect where at some point the body hits its capacity to improve. At the elite level of human performance, there seems to be some kind of biological performance ceiling—when you're near the top, there's only so much room left for improvement. Doing three things that in isolation might produce a 3 percent improvement in recovery doesn't add up to a 9 percent boost. What this suggests is that these different treatments may be tapping into the same finite reserve. Whatever it is they're doing, they seem to be accomplishing it via some shared mechanism.

David Martin suspects that this mechanism is the placebo effect. Martin is an Oregon native and endurance athlete (Nordic skiing was one of his first passions). He spent two decades working at the Australian Institute of Sport before the Philadelphia 76ers hired him in 2015 to serve as their director of performance research and development. Over the years, he's worked with elite athletes from numerous sports, including Cadel Evans, Australia's first winner of the Tour de France, and NBA center Joel Embiid. After decades of observing athletes and their habits, Martin has concluded that most popular recovery modalities work by exploiting the placebo effect. But he doesn't see that as a reason to dismiss them. On the contrary, he views it as an opportunity. This is real mojo, and instead of calling it the placebo effect he prefers the terms "anticipatory response" or "belief effects." He uses these alternative names, because people tend to dismiss the word *placebo* as a synonym for ineffective, when, in fact, these effects are real, and in some cases can be as powerful as many drugs. The difference is that you're gathering up your body's own resources to create the benefit.

The body's natural powers can be amazingly potent. In one study, oral surgery patients given a placebo had a 39 percent reduction in pain, but this placebo response disappeared when participants received a drug that blocks opioid receptors, which suggests that the placebo effect exploits the body's natural opioid system.[2] Brain imaging suggests that neurotransmitters like endorphins may also play a role. The takeaway, Martin says, is that expectations can create real biological effects.

It doesn't matter if there's science to back it up. If an athlete strongly believes that something works, the belief effect can overwhelm the real effect, Martin says. And the reverse is true too. If the athlete doesn't believe in the modality, its benefits will be diminished or even erased.

"We know that the brain and the body work together," says Luana Colloca, a researcher at the University of Maryland who studies placebos. The placebo effect doesn't just represent an expectation, but also a prediction about an event, she says. Once you've had a good prior experience with a drug, if the next time you think you're taking it you inadvertently ingest an inert substance instead, research shows you're likely to feel like it worked. Watching someone else respond well to a treatment can prod you to expect that it will work for you too. It's not just true for drugs. If you become convinced that a rival or teammate got a boost from a recovery modality, you can feel a real benefit, even if it's nothing but the equivalent of an inert drug. Belief is a powerful thing.

A friend recently asked me what I thought of icing. I asked

her what she would say if I told her it didn't help. "I wouldn't believe you," she said. She knew that it worked for her. It was an honest answer that explains why we cling to unproven and even debunked modalities even in the face of evidence that they don't work. We know what we believe and won't be dissuaded.

Some years ago, David Nieman, director of the Human Performance Lab at Appalachian State University, conducted a study on participants of the Western States Endurance Run, a 100-mile ultramarathon in California's Sierra Nevada mountains.[3] Among ultra runners, the use of ibuprofen to manage soreness and pain is so common that they call the drug "vitamin I." Nieman recruited a group of racers and asked half of them to forgo ibuprofen during the event, something many would-be participants were loath to do. Afterward, he compared pain and inflammation in runners who took ibuprofen during the race with those who didn't, and the results were startling. Ibuprofen failed to reduce muscle pain or soreness, and blood tests revealed that ibuprofen takers actually experienced greater levels of inflammation than those who eschewed the drug. "There is absolutely no reason for runners to be using ibuprofen," Nieman said.

The following year, Nieman returned to the Western States race and presented his findings to runners. Afterward, he asked them whether his study results would change their habits. The answer was a resounding no. "They really, really think it's helping. Even in the face of data showing that it doesn't help, they still use it." After I wrote about Nieman's study, I received emails from runners questioning the results. It didn't matter what the research had shown, these readers told me, because "I know it works for me."

Since Nieman's study came out in 2006, other research

has shown that using ibuprofen and other nonsteroidal anti-inflammatory drugs (NSAIDs) prior to exercise may actually impede tissue repair and delay the healing of bone, ligament, muscle, and tendon injuries.[4] And because inflammation appears to be a necessary step in a muscle's adaptation to exercise, some researchers theorize that taking anti-inflammatories might blunt some of the gains that you'd otherwise get from hard training, an idea supported by preliminary studies. The downsides of NSAIDs have received attention in the *New York Times* and *Washington Post* as well as sporting publications like *Runner's World* and specialty websites targeted at athletes.

Yet the belief that NSAIDs should help athletes perform with less pain remains ingrained. One runner I interviewed was hospitalized for a severe case of rhabdomyolysis (a serious kidney condition) after taking ibuprofen during an ultramarathon. Despite her experience, attributable in part to the twelve ibuprofen pills she popped during the 24-hour run, she continues to take ibuprofen while racing, albeit in lower doses. "Ibuprofen absolutely does work for me," she told me. She credited the drug with reducing pain and joint inflammation. The notion that a class of drugs whose name includes the word "anti-inflammatory" could actually *increase* inflammation strikes many runners as not just improbable, but impossible—even though that's exactly what research has shown.

Stretching is another practice that's so ingrained that people rarely question its benefits. It's an excellent placebo, because it's ritualized, it provides a sense of agency, and it feels like something's happening, which can reinforce an expectation that it's working. I know. I grew up stretching. We had a whole routine of stretches that we did before and after every cross-country and track practice when I was in high school. I don't recall ever getting much explanation. It was just what you did to keep your muscles loose and feeling good.

Although stretching has always been a popular standby for athletes, recently it's become trendy too, with a series of stretching studios popping up around places like New York City, Washington, D.C., and the San Francisco Bay Area offering personalized one-on-one stretching sessions with a trainer. The idea here is that stretching helps sore muscles relax and relieves tension in muscles that are tight from exercise or, for office jockeys, from staying too long in a fixed position. Stretching is also promoted as a way to ensure flexibility and good range of motion in the joints.

Too bad that reviews of the scientific literature on stretching have deemed stretching all but useless for recovery. A 2011 Cochrane review analyzed twelve studies of stretching—one of them a field study that involved more than 2,000 participants—and concluded that "the evidence from randomised studies suggests that muscle stretching, whether conducted before, after, or before and after exercise, does not produce clinically important reductions in delayed-onset muscle soreness in healthy adults." A more recent review concluded that static stretching might reduce running economy (a measure of how efficiently you run), doesn't reduce the duration or intensity of delayed-onset muscle soreness, and doesn't make a dent in injury risk either.[5] Sure, stretching may make it easier to touch your toes or show off other signs of flexibility, but that doesn't seem to translate into fewer injuries or less soreness.

I quit stretching many years ago, after searching for the scientifically validated benefits of the practice and finding that there weren't many. As far as I can tell, giving up stretching has made absolutely no difference, though it has saved me time.

Training partners sometimes question my decision, but I long ago gave up trying to convince adherents. When well-meaning friends try to rope me into their stretching routines, I fall back on the Bartleby reply—I would prefer not to. If my workout partners want to spend time on stretching rituals, I won't try to stop them. I long ago realized that I won't change their minds with studies.

I've noticed again and again that people are quick to reject evidence that contradicts their personal experience. For instance, in placebo studies, you might expect that volunteers who got the sham treatment would feel deceived, but that's not usually the case.[6] Part of what a placebo does is change how we appraise sensory input. And this appraisal can make a big difference. If you're primed to expect to feel less sore or fatigued after icing or a massage, you may perceive your soreness as lessened and you may *actually* feel less tired. When you think about it, many of the things that recovery modalities are supposed to do have a subjective or psychological component that can be harnessed for good.

At the 76ers facilities, players are offered a selection of four recovery stations and allowed to choose the one that's best for them. The usual choices are massage, ice bath, pneumatic compression boots, and a warm or hot tub (or a manual stretch from a therapist if they're on the road). Martin, the 76ers exercise scientist, takes a careful approach when presenting these options to players, telling them that the evidence that they work isn't definitive (and in some cases isn't very strong) but that some people find them helpful and maybe they will too.

Several studies have shown that placebos can be effective medicine even when people are told they're inert.[7] Harvard researcher Ted Kaptchuk says that while so-called open placebos don't hinge on deception, they do employ some sleight of hand. A doctor who uses an open placebo is like a magician. The trick is performed with full disclosure that it is, in fact, a trick, but it still requires a subtle form of deception to execute. For the placebo to work, the patient must suspend disbelief at the doctor's urging, which is pretty much exactly what Martin is doing with his players. Martin's gentle, confident manner and his reputation as a sports science genius probably help too, by earning players' trust.

By design, Martin offers athletes the opportunity to choose their recovery modality. "If they choose it, they back it 100 percent, but if I tell them they have to do it, it doesn't work," he says. He recalls a study done by researchers at the University of Toledo in Ohio in which volunteers were offered a selection of placebo creams for pain.[8] If they were shown the creams and then assigned one, the creams were deemed less effective than when they were offered the selection and allowed to choose. Something about making the decisions themselves enhanced the placebo effect.

Many popular recovery modalities strike me as a sort of pacifier. They won't actually resolve anything, but they give you something to do while you wait for nature to take its course. There's ample research to show that we humans hate waiting, and if you give us something to occupy us in the interim, we'll be much happier, even if nothing else has changed. For instance, when passengers at the Houston airport complained about long wait times at the baggage carousel, airport officials tried speeding up the process for transporting luggage.[9] The approach shortened the time, but still customers complained. It turned out that the walk from the gates to the baggage claim was very short so that even with expedited bag transport, people ended up spending almost 90 percent of the time after deplaning waiting for their luggage. The solution? They rerouted passengers so that it took six times longer to arrive at baggage claim. Complaints all but disappeared. The wait time itself hadn't changed, only how people spent it. Turns out, we're much happier doing something—anything—than we are waiting. Stretching or icing or foam rolling provides a sense of agency and engagement, which studies point to as active ingredients in placebos. "If something takes a lot of commitment on our part, or if it costs more to you, it seems to work better," says Tor Wager, a placebo researcher at the University of Colorado.

Experiments have identified some other characteristics that seem to give placebos power. Anything that gives the sense that the placebo is strong can make it seem like a genuine drug, which may explain why placebo shots are more effective than placebo pills, and placebo surgeries are even better. In studies, researchers sometimes give what they call "active" placebos, which have some kind of benign effect—a weird taste in your mouth or a tingling on your tongue, perhaps—to enhance the expectation that something is happening.

Many of these placebo-enhancing characteristics turn up in recovery modalities, which I've come to think of as falling into four categories. First is the "feels so good" bin. These are things that feel good while you're doing them, and perhaps afterward too, which even if it does nothing else, provides a valuable benefit in itself. Next is the "hurts so good" bin, with things like icing that are painful, and thus give the sense that they must be powerful (and therefore effective). Third is the "it's working, I can sense it" bin of active placebos, with things like cupping, which produce noticeable sensations and effects that aren't necessarily painful nor alluringly pleasant. Finally there's the "blinded by science" bin of things like infrared saunas, whose appeal comes from jargony scientific explanations that give them an aura of space-age power.

These categories are fluid—what falls into the "feels so good" bin for one person might be categorized as "hurts so good" or "it's working, I can sense it" by another—but the ultimate effect is the same. The modality has some quality that triggers an expectation that it will work.

———

Natalie Badowski Wu is an ultramarathoner and ER physician in San Jose, California. "We runners are kind of a type," she says. "We hate to do nothing. We're always finding ways to promote recovery, so you're doing something to feel like you're having an effect." She's noticed a proliferation of recovery modalities offered at ultra races. Massages, ice baths, and pneumatic compression devices have become so common, she says, that runners almost demand them now. "There's a lot of company-run tests where they say, oh yeah, there's a little bit of evidence with this," she says, but she wanted to see some stronger proof and studies that applied to runners. "Some of the studies will have people do a one-rep max on some weight exercise, but I'm like, what does that mean for running?"

Randomized, controlled trials are the gold standard in medicine, and Badowski Wu and some colleagues, including Martin Hoffman, the Sacramento VA Medical Center researcher who wrote hydration guidelines, figured they could apply that approach to recovery modalities too. So they designed and conducted a randomized, controlled trial of massage and pneumatic compression on participants in the 2015 Western States Endurance Run.[10] (The race has a unique program to fund and encourage scientific research on the event and its participants.) The runners were randomly split into three groups to get either massage, pneumatic compression treatment, or nothing after finishing the race.

Runners who received massage or a compression treatment reported reduced muscle pain and soreness immediately afterward, but these benefits were short-lived. The results showed no difference between the control group and the two treatment groups on measures of muscle pain and soreness or overall fatigue. Researchers also asked the runners to perform two 400-meter time trials before the race, and then again three and five days after the race, and here too there were no differences between groups.

Badowski Wu wonders whether the improvements they saw immediately after the treatments were influenced by an expectation effect. "You can't blind someone to these interventions—you know you're getting a massage or wearing the compression pants, and people think if something is being done, there must be some benefit." Some of the runners assigned to the squeezy pants were visibly excited, while some of those assigned to nothing were clearly bummed, she says. "I could tell that some people assigned to the control group were quite disappointed. Did that bias their scores lower? Is there not really a difference and it's just in our heads? Is it a self-fulfilling prophecy?" The group is planning some follow-up studies to further explore questions like these.

I asked Badowski Wu what *she* does for recovery. She says that when she ran track at Washington University in St. Louis she used to do ice baths, but quit after seeing research showing that while ice baths might be great in between her 400-meter heats at a meet, the baths seem detrimental to muscle repair over the long term. Now she's diligent about foam rolling. She also bought a pair of pneumatic compression pants a while back. "I still don't know if it helps me, but it feels so good afterward, and I do think there's some benefit to that."

There's a ritualization to some of these recovery modalities that shouldn't be overlooked. In a recent editorial, Jonas Bloch Thorlund from the University of South Denmark deconstructed why arthroscopic surgery for meniscal tears remains popular, despite compelling evidence that these procedures are essentially placebos, no better than sham surgery.[11] Thorlund notes that surgery represents a ritualistic activity that fosters expectations, much like the way shamans do. There's the journey to a healing place (the hospital), anointment with a purifying liquid (the presurgical skin prep), and an encounter with the masked healer. As I read this description, I felt a glimmer of recognition, thinking about my experiences visiting recovery centers. In each case, you're greeted by an empathetic caregiver who walks you through a series of rituals that require various forms of preparation and waiting. It makes me wonder how much power resides in the simple act of putting your trust in a healer and taking part in the ritual on offer.

A good example of this comes from cupping, a practice developed generations ago in China that involves placing glass suction cups along the body to draw up the skin (and create circular bruises). "It is novel and exotic, it is an ancient secret, and most of all, it leaves a mark behind to symbolize that 'something' happened," as Steve Magness has put it.[12]

Olympic skier Mikaela Shiffrin, the World Cup's second-youngest ever overall champion, swears by cupping. She says it helps her muscles "stay in line." She likes to get cupped when her back gets tight. "It feels like you're getting pinched really hard," she says, likening it to a kind of deep massage, but instead of pushing down, it pulls the fascia up, offering, she says, a different form of release. "My physio puts five cups in a row up my iliotibial band. It's very unpleasant when it's happening, but helps the muscle relax and chill out." For her, cupping falls into the "hurts so good" mind-set.

I've tried it myself, and would put cupping squarely in the "it's working, I can sense it!" category. It didn't really hurt, but it didn't feel good either. Mostly it felt weird. There could be no doubt that *something* was happening, but I couldn't convince myself that it was something good. My one session left only red spots and one minor bruise, but I wondered if the trick was that the bruises make you forget whatever pain you were trying to address in the first place.

After gold medalist swimmer Michael Phelps showed up at the pool sporting purple cupping marks on his shoulders and back at the Rio Olympics, James Hamblin, a writer at *The Atlantic* who is also an MD, wrote an article in the form of a plea: "Please, Michael Phelps, Stop Cupping."[13] Despite claims that cupping increases blood flow, "A bruise is a blood clot," Hamblin notes, "and clotted blood is definitionally not flowing." Beliefs about cupping's benefits "are multitudinous, limited only by the imagination. And, in terms of scientific evidence, substantiated only by the imagination," Hamblin writes, noting that there are no rigorous studies of the method. Like most recovery modalities, it's nearly impossible to have a blinded trial, because people can easily tell if they're getting the treatment or not.

I'm not sure that athletes really care about the scientific evidence on things like cupping. So often what athletes are really seeking when they turn to these expectation-inducing tools isn't physical, but psychological. They want to feel proactive and empowered. More than anything else, they're seeking confidence. They want to feel like they've done everything they can to take care of themselves and facilitate recovery.

———

You might expect that people who use recovery modalities, with all their scientific-sounding promises, have a performance edge, but that's not necessarily true. Athletes housed at the Olympic Training Center had nearly unlimited access to the recovery center, and performance technologist Bill Sands kept records of who used massage and how often. After the Olympics in 2008, he compared performance in Beijing with recovery center usage and found that nonmedalists used the recovery center massage services more than twice as much as the medalists. What this suggests to Sands is that the best athletes don't need all these recovery tools, at least not when they're on form. "It appears that recovery centers may be more ideal for athletes who aren't as good," he says. "The kids who went to the Olympic Games and won medals used the recovery center statistically less than those who used the recovery center a lot." He speculates that athletes "on the bubble," and struggling to make the team, are pulling out all the stops and possibly overextending themselves, and in that case they may need extra help. "If you have to really kill yourself to make the cutoff, you'll work your butt off much more than the kid that's more talented," he says.

Looking at all the time and effort that's required to put yourself through these recovery modalities, I have to wonder if an overemphasis on recovery tools could backfire if they mean that athletes spend even more time at the gym or training center, rather than relaxing at home.

There's a definite risk that recovery rituals can become their own source of stress, says Steve Magness, the running coach. "If you step back and think about it, what am I trying to do with recovery? I just worked out really hard, my entire body is essentially switched on, my nervous systems are firing, I've got all this adrenaline, all these hormones, all this damage. If I'm trying to get into 'recovery mode,' I'm essentially trying to switch that off," he says. "But if now all of a sudden I'm rushing to the cryosauna or I'm jumping into the ice bath (which is another stimulus that could produce some adrenaline) and if I'm worrying about spending the next 30 minutes going through a stretching and rolling routine, those are all active things where my mind is engaged." Instead of winding down, you're essentially extending the work day, and the end result may be the opposite of what's intended.

Of course, whether or not a recovery program becomes a new source of stress may depend on how an athlete views it. Runner Todd Straka estimates that he spends about 75 percent as much time working on his recovery as he does training, and he doesn't view this as a burden, but as a way to take care of himself. He belongs to FixtMovement, a recovery center in Boulder where members can drop in anytime to use their large array of recovery toys, like NormaTec boots and an infrared sauna. (They also provide a small beer cooler filled with a specially produced microbrew.)

Straka says it took him fifteen or twenty years to get good at running. He didn't run in high school or college, but when he moved to Boulder in 1993 at age twenty-six, he caught the running bug. After turning forty, he started finding some success in the age-group competitions, and eyeing his fiftieth birthday approaching, he realized that the American record in that age group for the mile run might be within reach. (In the summer of 2017, he came within one-and-a-half seconds of breaking the record.) His coach, Ric Rojas, helped him focus his training on the goal by putting him on more short speed work and a middle-distance-specific plan. He'd previously done marathons and half-marathons, but while he was training for the mile, his mileage dropped (down to 35 or 40 miles per week) and his intensity rose, along with his recovery needs.

Straka's standard routine at FixtMovement includes a half hour in the NormaTec boots, some time on an inversion table, and maybe 15 minutes on what he calls the "back rack"—a table with jade rollers on it that heat up and massage the back. "It feels amazing," he says. Straka only recently started using the infrared sauna, which he prefers over regular saunas, because it's noticeably drier and feels less intensely hot. The sauna is a new thing for him, and he hasn't decided whether it's helping. But as he puts it, "It's always nice to be warm."

———

I've come to think that different recovery modalities just represent variations on the same few approaches to recovery— soothing your muscles and body so you *feel* better (even if nothing is actually changing in a physiological sense), providing a ritual for taking care of yourself that gives you a sense of autonomy and self-efficacy (what many people think of as being proactive), and finally, creating a formalized way to stop everything else and help you focus on resting.

Perhaps the most crucial thing that recovery modalities do is give athletes a means to be deliberate about recovery. When cyclist Taylor Phinney went to his first Olympics in 2008 to compete in the men's 4,000-meter individual pursuit (just months after his first time riding in a velodrome), his coach Neal Henderson knew he would need some supervision. The son of two famed American cyclists, Phinney was an eighteen-year-old kid at the time, prone to distraction.[14] His room in the Olympic Village was located underneath the US women's gymnastics team, and (as he confessed to *ESPN the Magazine*) he developed a "massive crush" on gymnast Shawn Johnson, throwing Snickers bars up to her room and spending more than a little time pursuing her.[15] To encourage focus, Henderson put Phinney in compression boots. "NormaTecs were the perfect thing for Taylor, because I knew he was not going to be doing anything—just down and out. That was a good way to take up some time in his day," Henderson says.

Whether you're a teenage boy with an infatuation or a female CEO with a million demands on you, Henderson says that one of the most important benefits of tools like compression boots or massage is that they force people to take a time-out. "Sometime these modalities literally are just putting a stop to everything else for a little bit."

Lauren Fleshman, five-time NCAA champion and former national champion at 5,000 meters, has noticed this too. "A lot of things work because they just make you sit down for 15 to 20 minutes." Fleshman, who retired from professional racing in 2016 at age thirty-four, coaches a group of elite runners and heads Picky Bars, an energy bar company she founded in 2010. She still runs regularly and sometimes visits a recovery gym in Bend, Oregon, called Recharge Sport, which offers massage therapy, contrast baths, and loads of recovery tools. "They have a big sectional with recliners and NormaTec boots. People will be lined up for them," Fleshman says. She's a fan of contrast baths and the compression boots, and she gets a weekly massage. "I think it's really helpful to have a consistent body worker who knows what your body feels like. My massage therapist is really good at noticing, oh, your muscle here is tight or your left side is unusually tired." The massage is a way to check in on what's happening in her body, and honestly, "It just feels really good to lie still for an hour, or even just a half hour."

Athletes seem to thrive on rituals and even the most scientifically minded among us can be prone to superstitions. I'm a science nerd, but I think that's okay. On competition day, the most important thing an athlete needs is confidence. If a placebo ritual can provide a confidence boost, then (assuming it's causing no harm) why not let them have it? If I've learned anything about recovery, it's that the subjective sense of how it feels is the most important part.

Conclusion

During my eight years as a serious Nordic ski racer, I fell into the same pattern. I'd train hard in the preseason and enter the first few races in top form. Then something would happen and I'd tweak a muscle or come down with a debilitating cold or flu bug. Just as I was on the verge of accomplishing a big goal, my season would be shot. At the time, I chalked up my failures to bad luck—I just happened to fall sick or injured every time I was reaching my peak. But I've come to understand that just as I was able to get fit fast, I was also prone to overtraining. I needed less training than most athletes to reach and maintain peak conditioning, but I did not appreciate that I also needed more rest and recovery.

If there's one thing I wish I could go back and tell my younger self it would be this: learn to read your own body and pay attention to what it's telling you. My susceptibility to injury and overtraining limited my potential in the same way that my aerobic capacity, long limbs, and unusually fast response to training raised it. In the end, my fragility wasn't simply a matter of bad luck, it was the thing I needed to manage if I were to reach my athletic potential. Instead, I too often ignored my body's cries for rest and tried to will myself well. Whether it was a nagging fatigue that I wished I didn't have or the twinge in my hamstring that reappeared at inconvenient times, my knee-jerk response for too long was to cover my ears and say, nah-nah-nah, I don't hear you! Deep down, I knew that what my body really needed was a break, but my mind didn't want to accept this. I get antsy when I go too long without moving. But denying yourself needed recovery only digs you into a deeper hole, and I've come to realize that the proper way to answer my body's cries for rest isn't to push through, but to master the art of stillness.

My days of serious competition are behind me, and today I train for fitness and fun. I no longer aim to be the best. I strive to stay fit enough to be able to do the things I love to do—long trail runs through the high peaks, epic mountain bike rides, and lots of cross-country skiing. In my fourth decade, I've noticed myself needing a lot more rest to recuperate after a long or hard workout. It's not just me—science has confirmed that we require more recovery as we age. For example, research shows that delayed-onset muscle soreness takes longer to resolve in older athletes. Thomas M. Doering at Central Queensland University in Australia (he has since moved to Bond University) did a small study of triathletes in their fifties showing that rates of muscle protein synthesis in these masters athletes were lower than those of athletes in their twenties.[1] His research suggests that older athletes may have an "anabolic resistance" to protein that makes it harder for their bodies to convert protein into muscle, and that may help explain why it takes longer for them to repair exercise-induced muscle damage. (It's also an argument for keeping your protein intake up as you age.)

I used to lament my increasing need for recovery, but over time I've come to embrace it. I go hard when appropriate, but I also enjoy my newfound art of relaxation. I'm discovering that I enjoy exercising for pleasure and stress relief. Some days, a long walk with my dog is enough to fill my needs. I still love to get my heart rate up and push myself once in a while, but not every bout needs to become a programmed workout.

———

As of May 2018, Mike Fanelli had logged 109,126 miles of running in his life. "I've trained every single day, pretty much since October of 1970," he says. That's nearly half a century of running every day, and he doesn't just run. "I train, and I train to compete. Every single workout has a distinct purpose." He's not as fast as he once was—he ran a 2:25 marathon in 1980 and a 4:56 mile in 2006, at age fifty—but he's still competitive. He lives just north of San Francisco and started running as a high school freshman in Philadelphia. He started as a half-miler, and moved up to marathons (he won the San Francisco Rim Marathon twice) and even 100-mile races, but in recent years has returned to the track to focus on the 800 and 1,500 meters. "I've pretty much done it all," he says.

Before he gets out of bed each morning, Fanelli is already working on recovery by doing some trunk twists, light stretches, and foot exercises. Only then does he step into his shoes with special insoles. The very next thing he does is take a "wellness" formula out of a dropper. He also downs a collagen supplement, and to reduce inflammation he takes 900 mg of turmeric three to four times per day, including in the middle of the night when he gets up to pee. He's gotten weekly massages since 1978 and has also tried myofascial release, Rolfing (a kind of massage), and active release therapy. After high-intensity track workouts, he visits a cryotherapy chamber or uses NormaTec compression boots, and he applies a 10 percent ibuprofen topical solution directly to his aches and pains. (He orders it, ten to twenty tubes at a time, from a pharmacy in the UK.) He's used a phosphate buffer supplement called "Stim O Stam" for every hard workout and race since the 1970s, and says it markedly decreases his muscle soreness. He also takes "a lot" of vitamins. "I've got the most expensive pee in Sonoma County," he laughs. "Are vitamins hocus-pocus or do they work? Some are probably better than others." To top it all off, he takes a "really hot" Epsom salt bath (with three to five pounds of salt) every night for 20 minutes while enjoying a nice glass of "well-deserved wine."

Fanelli runs about 40 miles per week these days and acknowledges that he spends "three times as much time" doing all these recovery things, but as a real estate agent in the San Francisco area he can afford it. "I believe that fitness takes place in the recovery phase," he says, so he "goes a little overboard" on recovery. "I believe in bombarding yourself with care."

I assume that with all this self-care, he must be really healthy and injury-free. "Oh no!" he laughs. "My feet are destroyed. Both are due for surgery. I'm in continual management mode. My right foot never bends the way it's supposed to bend." For nine years, he says, "I've been in excruciating pain, like someone was jamming a red hot poker into the top of my foot. The other foot has no fatty pad on it, so it's nerve on bone. The answer is no. I'm always injured and have stuff going on. My knees and hips are okay, but my feet are worn out, done."

I don't ask why he puts himself through all this just to keep running. I'm a runner too. I understand. "I'd never recommend that anybody go through the crap I go through," he says, but it works for him. He insists he's not as obsessed as he might sound. "I swear, I have a life—a beautiful wife and a fantastic dog, a Vizsla, who's a fantastic running dog."

Camille Herron set a world record the first time she finished a 100-mile run. In November 2017, the then-35-year-old was the first runner across the finish line at the Tunnel Hill 100 in southern Illinois. Her time of 12 hours, 42 minutes, and 39 seconds—a 7:38 per mile pace—earned her a world record for the fastest recorded finish (for men or women) on a certified 100-mile course.

Instead of meticulously structuring a recovery program in advance, Herron used intuition as her guide. She recovers by *feel*. "I am really in tune with my body, and I pay attention to what I'm feeling," she says. "After a marathon or ultra, it's usually about getting as many calories as I can. It can be hard to get enough quality proteins, and sometimes I just crave a cheeseburger so that's what I eat." She doesn't count calories or stress about finding the absolute perfect recovery fuel. Instead, she looks around at what's available, and gives herself whatever sounds most appealing. After one recent ultramarathon, she found a seafood restaurant at the airport and filled up on soft-shell crab, French fries, and a beer. "Oh, that was so good," she recalls. After a long race, she does a reverse taper to work her way back into training and does only as much mileage as her body can tolerate without feeling drained.

Herron's overarching theory of recovery boils down to this: keep it simple and don't sweat the small stuff. She focuses on the big picture—how are my overall stress levels? Am I feeling invigorated or drained by my training? When a blood test for a bladder problem came back with signs that her stress hormones were elevated, she recognized it as a red flag. "I realized that I needed to change my mental outlook," she says. Running had become a source of stress, so she ditched her GPS watch and

Strava and all the high-tech trackers and went back to a paper training log. "I needed to get back in touch with the joy of running," she says. "When I did that, I had one of the best training cycles I'd had in years." It wasn't compression boots or a special drink that turned her around (though she's used those sometimes). Instead, her ability to read her body and identify and eliminate unnecessary sources of stress was the key to her recovery and health. She enjoys a beer after a hard run, and sometimes even during a long event (at mile 80 of her record-setting 100-mile run, for instance). Whether it's the beer itself or the ritual and relaxation that it represents doesn't matter. It relaxes her, and that's what counts.

———

Mike and Camille are two very accomplished master athletes representing two polar opposite approaches to recovery. Which is the right one? I'm not sure there's a right answer. I think of aging and recovery like that wait at the Houston airport baggage claim. You can't change that recovery takes more time with age. Your only choice is about what you'll do in the meantime. Whether you stretch or pop ibuprofen or just say to hell with it is really a matter of temperament. You can be like Mike and grab on to every new trick to give yourself the confidence that you've done everything you possibly can, or you can be like Camille and keep it simple with moderation and caution.

As for me, I understand Mike's over-the-top rejuvenation program, and I admire his healthy respect for the importance of recovery. But count me in Camille's camp. Give me my morning walk, some meditation or floating, late mornings in bed, and maybe a bit of massage here and there, and then let every day end with a glass of wine and a sunset.

ACKNOWLEDGMENTS

When Matt Weiland first approached me about writing a book on exercise recovery, I stopped for a moment to consider whether I would find enough interesting material to fill several hundred pages. A year later, I wondered how I could possibly fit all my reporting into a single tome.

It wasn't easy, and a complete list of everyone who helped would surpass the allotted space. I'm grateful to each of the 223 people I interviewed during my research. Much of my thinking about recovery began at the Colorado Mesa University exercise lab. Gig Leadbetter, Gerry Smith, Michael Reeder, and Brent Alumbaugh were particularly helpful in the early stages of my research. No one knows more about recovery than Shona Halson, and she spent many hours answering my questions. David Martin, Kristen Dieffenbach, David Nieman, Stuart Phillips, Matt Dixon, and Neal Henderson were also especially helpful.

David Epstein and Alex Hutchinson were first-class companions on my journey through the various stages of book-writing angst. I'm grateful for insightful discussions about sports science with them and with Amby Burfoot, Mike Joyner, Steve Magness, Brad Stulberg, Jonathan Wai, and the late, great Terry Laughlin.

I can't imagine a more hardworking, effective, and supportive agent than Alice Martell. Her phone calls always lift my spirits. When I grow up, I hope to become the amazing writer she believes me to be.

Working with Matt Weiland has been one of the greatest fortunes and pleasures of my career. He's a thoughtful and skilled editor, but his superpower is his ability to make a writer feel sane and supported. He and his colleagues at Norton have been incredible partners in this endeavor. Remy Cawley made sure that each round of Matt's handwritten edits arrived at my doorstep on time. Zarina Patwa helped get the book across the finish line. My thanks also go to copyeditor Gary Von Euer and managing editor Rebecca Homiski. Will Scarlett made sure the book found an audience, and Steve Attardo went the extra mile to design a perfect cover.

Rosemerry Wahtola Trommer is both my BFF and creative muse. I completed the first draft of this manuscript at her riverside studio, which she calls "the barn," but will forever feel to me like a writing spa. Paolo Bacigalupi's friendship has been a sustaining force in my writing life. He threw me a lifeline when I couldn't figure out how to begin writing this book. Helen Fields provided many friendly nudges, and her framed sketch of me working in the "writer's cone of shame" sat on my desk for the duration. My sister, Jill Friesen, proved by example that it's possible to complete a difficult extracurricular task while holding down a job.

Anna Barry-Jester fed me almost daily affirmations and was always willing to listen to my ramblings. A woman of a million talents, she also took my author photo. Chad Matlin has been a steadfast supporter of this book from the start. His contention that a book is simply an "idea vehicle" proved surprisingly useful in the project's early stages. Nate Silver offered helpful advice on book writing, and Blythe Terrell cut me a little slack when I needed it in the final push to finish the manuscript. Working with Maggie Koerth-Baker and my other FiveThirtyEight colleagues has pushed me to become better in every way.

Laura Helmuth convinced me I was on track when I wasn't sure where I was going with this, and Farai Chideya offered helpful advice on how to balance book writing with everything else. Ann Finkbeiner and Richard Panek gave sound book-writing advice, and Erik Vance offered helpful ideas about placebos.

Amby Burfoot, Siri Carpenter, David Epstein, and Alex Hutchinson offered feedback on select chapters. Jonathan Dugas, Shona Halson, Mike Joyner, Catherine Price, Kristin Sainani, and Derek van Westrum read sections of the manuscript for accuracy. Dhrumil Mehta offered crucial help digging up journal articles. Meral Agish provided research help, and Lexi Pandell did the fact-checking. Remaining errors are all mine.

Special thanks to Ben Casselman for helping me find the jelly doughnut episode of "Ring of Fire" on YouTube. Clark Sheehan confirmed the identity of the rider Alex Stieda, who was featured in the episode.

Finally, my deepest thanks go to my husband, Dave. He taught me how to relax and unwind, and I wake up every morning thankful for the beautiful sanctuary of serenity he has built for us on our farm. A person writing a book makes a terrible spouse, but he never complained and willingly took on all the household duties I neglected as I was deep into writing. I couldn't possibly love anyone more than I love him.

NOTES

n the course of researching this book, I read hundreds of research papers and conducted more than 250 interviews. Unless otherwise noted, direct quotes in the text come from these interviews. The citations here reflect studies, papers, and interviews directly mentioned in the text, but for every direct citation there were usually several more sources that informed my thinking.

Some of my most insightful discussions took place in the hallways and meeting rooms at the 2016 American College of Sports Medicine meeting in Boston, and at the 2015, 2016, and 2017 MIT Sloan Sports Analytics conferences.

Introduction

1. Stoudemire posted the selfie on October 15, 2014. https://www.instagram .com/p/uLZKs5qA9_/.
2. Andy Hall, "Alipour's Stoudemire SportsCenter Sit-Down Spills Over," February 12, 2015, ESPN.com, https://www.espnfrontrow.com/2015/02/ alipours-stoudemire-sportscenter-sit-down-spills-over/.
3. Stoudemire's Instagram photo: https://www.instagram.com/p/uLZKs5qA9_/; his interview with ESPN's Sam Alipour: http://www.espnfrontrow.com/ 2015/02/alipours-stoudemire-sportscenter-sit-down-spills-over/.

1: Just-So Science

1. Ben Crair and Andrew Kehfeb, "German Olympians Drink a Lot of (Nonalcoholic) Beer, and Win a Lot of Gold Medals," February 19, 2018, *New York Times*, accessed February 24, 2018, https://www.nytimes.com/2018/02/19/sports/olympics/germany-olympics-beer.html.

2. Martin Pöchmüller, Lukas Schwingshackl, Paolo C. Colombani, and Georg Hoffmann, "A Systematic Review and Meta-Analysis of Carbohydrate Benefits Associated with Randomized Controlled Competition-Based Performance Trials," *Journal of the International Society of Sports Nutrition* 14 (2016): 1–12, https://doi.10.1186/s12970-016-0139-6.

3. Richard Feynman made his famous comment about how easily people fool themselves in a 1974 commencement address at Caltech.

4. John P. A. Ioannidis, "Why Most Discovered True Associations Are Inflated," *Epidemiology* 19, no. 5 (2008): 640–48, https://doi.10.1097/EDE.0b013e31818131e7.

5. Marjan Bakker, Annette van Dijk, and Jelte M. Wicherts, "The Rules of the Game Called Psychological Science," *Perspectives on Psychological Science* 7, no. 6 (2008): 543–54, https://doi.10.1177/1745691612459060.

6. Evelyn B. Parr, Donny M. Camera, José L. Areta, Louise M. Burke, Stuart M. Phillips, John A. Hawley, Vernon G. Coffey, et al., "Alcohol Ingestion Impairs Maximal Post-Exercise Rates of Myofibrillar Protein Synthesis Following a Single Bout of Concurrent Training," ed. Stephen E. Always, *PLoS ONE* 9, no. 2 (2014), Public Library of Science: e88384, https://doi.10.1371/journal.pone.0088384.

7. Matthew J. Barnes, Toby Mündel, and Stephen R. Stannard, "A Low Dose of Alcohol Does Not Impact Skeletal Muscle Performance After Exercise-Induced Muscle Damage," *European Journal of Applied Physiology* 111, no. 4 (2011): 725–29, https://doi.10.1007/s00421-010-1655-8; Matthew J. Barnes, "Alcohol: Impact on Sports Performance and Recovery in Male Athletes," *Sports Medicine* 44, no. 7 (2014): 909–19, https://doi.10.1007/s40279-014-0192-8.

2: Be Like Mike

1. "Be Like Mike Gatorade Commercial (ORIGINAL)," available on YouTube, 1:00, posted October 23, 2006, https://www.youtube.com/watch?v=boAGiq9j_Ak.

2. There are conflicting accounts of exactly what question sparked the research that led to Gatorade and who it was that asked. An official history published on the Cade Museum for Creativity & Invention website says that "Gatorade was the result of an offhand question posed in 1965 by former University of Florida linebacker Dwayne Douglas to Dr. J Robert Cade, a professor of renal medicine. 'Why don't football players ever urinate during a game?'" (accessed May 13, 2018, https://www.cademuseum.org/history.html). According to a history of Gatorade published on the company's website in 2017, "In early summer of 1965, a University of Florida assistant coach sat down with a team of university physicians and asked them to determine why so many of his players were being affected by heat and heat related illnesses." Both sources say that the researchers involved in developing the drink were Dr. Robert Cade, Dr. Dana Shires, Dr. H. James Free, and Dr. Alejandro de Quesada. The Gatorade official history is archived at https://web.archive.org/web/20170116204425/http://www.gatorade.com/company/heritage.

3. Darren Rovell, *First in Thirst: How Gatorade Turned the Science of Sweat into a Cultural Phenomenon* (New York: AMACOM, 2006).

4. Rovell, *First in Thirst*.

5. Ben Desbrow, a sports dietitian at Griffith University in Australia, had an epiphany about beer: people voluntarily drink a large volume of it compared to other beverages. It dawned on him that if he could harness the drinkability and likability of beer in a beverage that had a suitable nutrient profile for recovery, then maybe people would drink it in amounts that would help them retain more fluid. So he started tinkering with what was in the brew, taking commercial beers and manipulating the alcohol and sodium content. In one of the initial studies, they experimented with adding sodium to the beer. "It was disgusting," Desbrow says. "It tasted like you'd gone fishing and knocked the beer overboard and then applied the three-second rule." He's since been working to tweak the formula to make it a more palatable recovery beer, but the exercise seems a bit pointless, given that beer drinkers can easily get their salt from snacks that are already commonly consumed with brewskis.

6. Nadia Campagnolo, Elizaveta Iudakhina, Christopher Irwin, Matthew Schubert, Gregory R. Cox, Michael Leveritt, and Ben Desbrow, "Fluid, Energy and Nutrient Recovery via Ad Libitum Intake of Different Fluids and Food," *Physiology & Behavior* 171 (2017): 228–35, https://doi.10.1016/j.physbeh.2017.01.009.

7. Rovell, *First in Thirst.*

8. Bob Murray, "Preventing Dehydration: Sports Drinks or Water," May 20, 2005, accessed January 13, 2018, https://www.iahsaa.org/Sports_ Medicine_Wellness/Heat/GSSI-Preventing_Dehydration_Sports_Drinks_ or_Water.pdf.

9. Timothy David Noakes and Dale B. Speedy, "Lobbyists for the Sportsdrink Industry: An Example of the Rise of 'Contrarianism' in Modern Scientific Debate," *British Journal of Sports Medicine* 41, no. 2 (2017): 107–9.

10. Noakes and Speedy, "Lobbyists for the Sportsdrink Industry," 107–9.

11. V. A. Convertino, L. E. Armstrong, E. F. Coyle, G. W. Mack, M. N. Sawka, L. C. Senay, and W. M. Sherman, "American College of Sports Medicine Position Stand. Exercise and Fluid Replacement," *Medicine and Science in Sports and Exercise* 28, no. 1 (1996): i–vii, http://www.ncbi.nlm.nih .gov/pubmed/9303999.

12. Carl Heneghan and David Nunan, "Forty Years of Sports Performance Research and Little Insight Gained: Sports Drinks," *British Medical Journal* 345 (2012), https://doi.10.1136/bmj.e4797.

13. As other researchers evaluating clinical trials had done, Heneghan's group defined as "small" those studies with fewer than one hundred participants in each group.

14. Deborah Cohen, "The Truth about Sports Drinks," *British Medical Journal* 345 (2012): 20–25, https://doi.10.1136/bmj.e4737.

15. Amby Burfoot recounted this experience to me in an interview. He also writes about it in a preface to Timothy Noakes's *Waterlogged: The Serious Problem of Overhydration in Endurance Sports* (Champaign, IL: Human Kinetics, 2012), loc. 2456–2458, Kindle.

16. David L. Costill, Walter Kammer, and Ann Fisher, "Fluid Ingestion During Distance Running," *Archives of Environmental Health* 21, no. 4 (1970): 520– 25, http://www.tandfonline.com/doi/abs/10.1080/00039896.1970.10667282.

17. Michael N. Sawka, Louise M. Burke, E. Randy Eichner, Ronald J. Maughan, Scott J. Montain, and Nina S. Stachenfeld, "Exercise and Fluid Replacement," *Medicine and Science in Sports and Exercise* 39, no. 2 (2007): 377– 90, https://doi.10.1249/mss.0b013e31802ca597.

18. Yannis Pitsiladis and Lukas Beis, "To Drink or Not to Drink Recommendations: The Evidence," *British Medical Journal (Clinical Research Ed.)* 345 (2012): e4868, https://doi.10.1136/bmj.e4868.

19. Noakes is perhaps most famous for his theories about exercise fatigue and has made a career out of pushing against conventional scientific wisdom, some say to his own detriment. In 2017, the Health Professions Council of South Africa cleared him of a charge of professional misconduct that had been brought by the Association of Dietetics, which had complained about advice he'd given on Twitter telling a mother to feed her baby a low-carb, high-fat diet—an eating plan that's the subject of his latest crusade.

20. T. D. Noakes and D. B. Speedy, "Case Proven: Exercise Associated Hyponatraemia Is Due to Overdrinking. So Why Did It Take 20 Years before the Original Evidence Was Accepted?" *British Journal of Sports Medicine* 40, no. 7 (2006): 567–72, https://doi.10.1136/bjsm.2005.020354.

21. T. D. Noakes, N. Goodwin, B. L. Rayner, T. Branken, and R. K. Taylor, "Water Intoxication: A Possible Complication during Endurance Exercise," *Medicine and Science in Sports and Exercise* 17, no. 3 (1985): 370–75.

22. Noakes, *Waterlogged*.

23. Tyler Frizzell, "Hyponatremia and Ultramarathon Running," *Journal of the American Medical Association* 255, no. 6 (1986): 772–74.

24. Noakes and Speedy, "Case Proven: Exercise Associated Hyponatraemia."

25. Eric Zorn, "Runner's Demise Sheds Light on Deadly Myth," *Chicago Tribune*, October 11, 1999.

26. Christopher S. D. Almond, Andrew Y. Shin, Elizabeth B. Fortescue, Rebekah C. Mannix, et al., "Hyponatremia among Runners in the Boston Marathon," *New England Journal of Medicine* 352, no. 15 (2005): 1550–56, https://doi.10.1056/NEJMoa043901.

27. Matthias Danz, Klaus Pottgen, Philip Tonjes, Jochen Hinkelbein, and Stefan Braunecker, "Hyponatremia among Triathletes in the Ironman European Championship," *New England Journal of Medicine* 374, no. 10 (2016), https://doi.10.1056/NEJMc1514211.

28. M. H. Rosner and J. Kirven, "Exercise-Associated Hyponatremia," *Clinical Journal of the American Society of Nephrology* 2, no. 1 (2006): 151–61, https://doi.10.2215/CJN.02730806.

29. William O. Roberts, "Exertional Heat Stroke During a Cool Weather Marathon," *Medicine and Science in Sports and Exercise* 38, no. 7 (2006): 1197–1203, https://doi.10.1249/01.mss.0000227302.80783.0f.

30. Tamara Hew-Butler, Valentina Loi, Antonello Pani, and Mitchell H. Rosner, "Exercise-Associated Hyponatremia: 2017 Update," *Frontiers in*

Medicine 4 (2017), https://doi.10.3389/fmed.2017.00021; Tamara Hew-Butler, Mitchell H. Rosner, Sandra Fowkes-Godek, Jonathan P. Dugas, et al., "Statement of the Third International Exercise-Associated Hyponatremia Consensus Development Conference, Carlsbad, California, 2015," *Clinical Journal of Sport Medicine* 25, no. 4 (2015): 303–20, https://doi.10.1097/JSM.0000000000000221.

31. C. Heneghan, P. Gill, B. O'Neill, D. Lasserson, M. Thake, M. Thompson, and J. Howick, "Mythbusting Sports and Exercise Products," *British Medical Journal* 345 (2012), https://doi.10.1136/bmj.e4848.

32. Ronald J. Maughan, Susan M. Shirreffs, and John B. Leiper, "Errors in the Estimation of Hydration Status from Changes in Body Mass," *Journal of Sports Sciences* 25, no. 7 (2007): 797–804, https://doi.10.1080/02640410600875143.

33. Martin D. Hoffman, "The Basics of Proper Hydration During Prolonged Exercise," posted on the Ultra Sports Foundation website, 2017, accessed January 2018, http://ultrasportsscience.us/wp-content/uploads/2017/07/The-Basics-of-Proper-Hydration.pdf.

34. Hayden Bird, "Medical Experts Offer Response to "TB12 Method" Claim about Avoiding Sunburn through Hydration," September 29, 2017, Boston.com, accessed August 7, 2018, https://www.boston.com/sports/new-england-patriots/2017/09/29/tb12-method-sunburn-prevention-hydration-claim-doctor-response; Vivian Manning-Schaffel, "Tom Brady Says This Trick Prevents Sunburns. Science Says Otherwise," September 27, 2017, NBC, accessed August 7, 2018, https://www.nbcnews.com/better/health/tom-brady-s-drinking-water-prevents-sunburn-claim-fake-news-ncna805116.

3: The Perfect Fuel

1. The jelly doughnut segment appeared on episode two ("Change") of the television program *The Ring of Truth*. The six-part series aired on PBS in 1987. The show was hosted by Philip Morrison, a scientist on the Manhattan Project, which developed the first nuclear weapon. He later became a vocal advocate for arms control. He died in April 2005 at age 89. As of February 15, 2018, the episode was available on YouTube at https://www.youtube.com/watch?v=Nk8CQNThbc0.

2. In 1986, Alex Stieda, a Canadian from Vancouver, became the first North American to wear the yellow leader's jersey in the Tour de France.

3. A press release published on November 9, 1998, touted a study presented at the American College of Sports Medicine Mid-Atlantic meeting and conducted by Michael Williams and Drs. John Ivy and Peter Raven.

4. J. L. Ivy, A. L. Katz, C. L. Cutler, W. M. Sherman, and E. F. Coyle. "Muscle Glycogen Synthesis after Exercise: Effect of Time of Carbohydrate Ingestion." *Journal of Applied Physiology (Bethesda, Md.: 1985)* 64, no. 4 (1988): 1480–85, doi:10.1152/jappl.1988.64.4.1480.

5. John Ivy and Robert Portman, *Nutrient Timing: The Future of Sports Nutrition* (Laguna Beach, CA: Basic Health Publications, 2004).

6. According to a biography listed on the PacificHealth Laboratories company website, Robert Portman cofounded M.E.D. Communications in 1974, and "the company grew to one of the largest medical agencies in the US." In 1993, he created C&M Advertising, "a consumer/medical agency with billings in excess of $100 million." He also holds twelve patents for nutritional interventions targeting sports performance, appetite, and diabetes; http://www.pacifichealthlabs.com/investor-center-directors/.

7. Michael Goodwin, "Blood-Doping Unethical, U.S. Olympic Official Says," *New York Times*, January 13, 1985, http://www.nytimes.com/1985/01/13/sports/blood-doping-unethical-us-olympic-official-says.html.

8. Paul Roberts, "Ed Burke's Got a Rocket in His Pita Pocket," *Outside*, May 1, 2001, https://www.outsideonline.com/1888016/ed-burkes-got-rocket-his-pita-pocket.

9. Alan Albert Aragon, Brad Jon Schoenfeld, C. Kerksick, T. Harvey, J. Stout, B. Campbell, C. Wilborn, et al., "Nutrient Timing Revisited: Is There a Post-Exercise Anabolic Window?" *Journal of the International Society of Sports Nutrition* 10, no. 1 (2013): 5, https://doi.10.1186/1550-2783-10-5.

10. Brad Jon Schoenfeld, Alan Aragon, Colin Wilborn, Stacie L. Urbina, Sara E. Hayward, and James Krieger, "Pre- versus Post-Exercise Protein Intake Has Similar Effects on Muscular Adaptations," *PeerJ—The Journal of Life and Environmental Sciences* 5 (2017): e2825, https://doi.10.7717/peerj.2825.

11. Schoenfeld and his colleagues published a meta-analysis on studies of protein timing on muscle adaptations and concluded that benefits like improved muscle building that were attributed to the timing of protein intake were instead probably due to the increase in protein consumed. Some of the studies that pointed to a metabolic window had methodological problems, such as comparing protein to a placebo, instead of comparing protein consumed at one time versus another. Other studies were too small to produce meaningful answers. Brad Jon Schoenfeld, Alan Albert

Aragon, and James W. Krieger, "The Effect of Protein Timing on Muscle Strength and Hypertrophy: A Meta-Analysis," *Journal of the International Society of Sports Nutrition* 10, no. 1 (2013): 53, https://doi.10.1186/1550-2783-10-53.

12. Michael J. Cramer, Charles L. Dumke, Walter S. Hailes, John S. Cuddy, and Brent C. Ruby, "Postexercise Glycogen Recovery and Exercise Performance Is Not Significantly Different between Fast Food and Sport Supplements," *International Journal of Sport Nutrition and Exercise Metabolism* 25, no. 5 (2015): 448–55, https://doi.10.1123/ijsnem.2014-0230.

13. Usain Bolt, *Faster Than Lightning: My Autobiography* (New York: Harper-Sport, 2013).

14. In January 2017, Bolt and his teammates were stripped of their 2008 Olympic gold medals in the 4x100 relay after Bolt's teammate Nesta Carter failed a drug test carried out on stored samples.

15. Baxter Holmes, "The NBA's Secret Addiction," *ESPN the Magazine*, March 27, 2017, http://www.espn.com/espn/feature/story/_/page/presents18931717/the-nba-secret-addiction.

16. Patrik Sörqvist, Daniel Hedblom, Mattias Holmgren, Andreas Haga, Linda Langeborg, Anatole Nöstl, and Jonas Kågström, "Who Needs Cream and Sugar When There Is Eco-Labeling? Taste and Willingness to Pay for 'Eco-Friendly' Coffee," ed. Amanda Bruce, *PLoS ONE* 8, no. 12 (2013): Library of Science: e80719, https://doi.10.1371/journal.pone.0080719.

17. E. Cockburn, E. Stevenson, P. R. Hayes, P. Robson-Ansley, and G. Howatson, "Effect of Milk-Based Carbohydrate-Protein Supplement Timing on the Attenuation of Exercise-Induced Muscle Damage," *Applied Physiology, Nutrition, and Metabolism* 35, no. 3 (2010): 270–77, https://doi.10.1139/H10-017; Kelly Pritchett and Robert Pritchett, "Chocolate Milk: A Post-Exercise Recovery Beverage for Endurance Sports," *Medicine and Sport Science*, 59 (2012): 127–34, https://doi.10.1159/000341954; Jason R. Karp, Jeanne D. Johnston, Sandra Tecklenburg, Timothy D. Mickleborough, Alyce D. Fly, and Joel M. Stager, "Chocolate Milk as a Post-Exercise Recovery Aid," *International Journal of Sport Nutrition and Exercise Metabolism* 16, no. 1 (2006): 78–91, https://doi.10.1097/00005768-200405001-00600.

18. M. P. McHugh Connolly and O. Padilla-Zakour, "Efficacy of a Tart Cherry Juice Blend in Preventing the Symptoms of Muscle Damage," *British Jour-*

nal of Sports Medicine 22, no. 4 (2006): 679–83, https://doi.10.1136/bjsm
.2005.025429.

19. Mayur K. Ranchordas, David Rogerson, Hora Soltani, and Joseph T.
Costello, "Antioxidants for Preventing and Reducing Muscle Soreness
after Exercise," *Cochrane Database of Systematic Reviews* (December
2017), https://doi.10.1002/14651858.CD009789.pub2.

20. Flanagan finished third in the 10,000 meters at the 2008 Beijing Olympics, but in August 2017, the International Olympic Committee upgraded
her medal to a silver after disqualifying second-place finisher Elvan Abey-
legesse of Turkey for failing a drug test. US Olympic Committee statement,
"Distance Runner Shalane Flanagan Upgraded to Silver Medal in 10,000-
Meter for Olympic Games Beijing 2008," accessed May 13, 2018, https://
www.teamusa.org/News/2017/August/21/Distance-Runner-Shalane-
Flanagan-Upgraded-To-Silver-Medal--In-10000-Meter-For-Beijing-2008.

21. Margo Mountjoy, Jorunn Sundgot-Borgen, Louise Burke, Susan Carter,
Naama Constantini, Constance Lebrun, Nanna Meyer, et al., "The IOC
Consensus Statement: Beyond the Female Athlete Triad—Relative Energy
Deficiency in Sport (RED-S)," *British Journal of Sports Medicine* 48, no. 7
(2014): 491–97, https://doi.10.1136/bjsports-2014-093502.

22. Asker E. Jeukendrup, "Periodized Nutrition for Athletes," *Sports Medicine* 47, no. S1 (2017): S51–63, https://doi.10.1007/s40279-017-0694-2.

4: The Cold War

1. James shared this series of ice bath images on his Instagram account on
October 2, 2013, https://www.instagram.com/p/e_ZmReCTJJ/.

2. Gabe Mirkin and Marshall Hoffman, *The Sports Medicine Book* (Little,
Brown, 1978).

3. Edward Swift Dunster, James Bradbridge Hunter, Frank Pierce Foster,
Charles Euchariste de Medicis Sajous, Gregory Stragnell, Henry J. Klaun-
berg, and Félix Martí-Ibáñez, *International Record of Medicine and General Practice Clinics, Volume 83* (MD Publications, 1906).

4. *The Inertia* posted the photo of Slater sitting in an ice bath on January 9,
2017, https://www.instagram.com/p/BPDjcTQBcQs/.

5. Austin Scaggs, "Madonna Looks Back: The Rolling Stone Interview," *Rolling Stone*, October 29, 2009.

6. Gabe Mirkin, "Why Ice Delays Recovery," September 16, 2015, Dr.Mirkin .com, accessed January 2018, http://www.drmirkin.com/fitness/why-ice-delays-recovery.html.

7. The standalone museum closed in 2002, but it still exists as the "Questionable Medical Device" collection at the Science Museum of Minnesota. "Museum of Quackery and Medical Frauds," Atlas Obscura, accessed May 10, 2018, https://www.atlasobscura.com/places/museum-quackery.

8. Motoi Yamane, Hiroyasu Teruya, Masataka Nakano, Ryuji Ogai, Norikazu Ohnishi, and Mitsuo Kosaka, "Post-Exercise Leg and Forearm Flexor Muscle Cooling in Humans Attenuates Endurance and Resistance Training Effects on Muscle Performance and on Circulatory Adaptation," *European Journal of Applied Physiology* 96, no. 5 (2006): 572–80, https://doi .10.1007/s00421-005-0095-3.

9. Ching-Yu Tseng, Jo-Ping Lee, Yung-Shen Tsai, Shin-Da Lee, Chung-Lan Kao, Te-Chih Liu, Cheng-Hsiu Lai, M. Brennan Harris, and Chia-Hua Kuo, "Topical Cooling (Icing) Delays Recovery From Eccentric Exercise-Induced Muscle Damage," *Journal of Strength and Conditioning Research* 27, no. 5 (2013): 1354–61, https://doi.10.1519/JSC.0b013e318267a22c.

10. Llion A. Roberts, Truls Raastad, James F. Markworth, Vandre C. Figueiredo, Ingrid M. Egner, Anthony Shield, David Cameron-Smith, Jeff S. Coombes, and Jonathan M. Peake, "Post-Exercise Cold Water Immersion Attenuates Acute Anabolic Signalling and Long-Term Adaptations in Muscle to Strength Training," *Journal of Physiology* 593, no. 18 (2015): 4285–4301, https://doi.10.1113/JP270570.

11. As every CrossFit aficionado knows, WOD stands for "workout of the day."

12. "Icing Muscles Information," YouTube video posted by Kelly Starrett on July 19, 2012, https://www.youtube.com/watch?v=0UmJVgEWZu4.

13. Kelly Starrett, "People, We've Got to Stop Icing Injuries. We Were Wrong, Sooo Wrong," Daily M/WOD (blog), accessed February 15, 2018, https:// www.mobilitywod.com/propreview/people-weve-got-to-stop-icing-injuries-we-were-wrong-sooo-wrong-community-video/.

14. Jeff Bercovici, "How This Fitness Entrepreneur Won Over Blake Griffin and LeBron James," *Inc.*, April 2015.

15. In August of 2018, Shona Halson left the Australian Institute of Sport to become an associate professor in the School of Behavioural and Health Sciences at Australian Catholic University.

16. J. Leeder, C. Gissane, K. van Someren, W. Gregson, and G. Howatson,

"Cold Water Immersion and Recovery from Strenuous Exercise: A Meta-Analysis," *British Journal of Sports Medicine* 46 (2012): 233–40, https://doi.10.1136/bjsports-2011-090061.

17. François Bieuzen, Chris M. Bleakley, and Joseph Thomas Costello, "Contrast Water Therapy and Exercise Induced Muscle Damage: A Systematic Review and Meta-Analysis," *PLoS ONE* 8, no. 4 (2013), https://doi.10.1371/journal.pone.0062356.

18. James R. Broatch, Aaron Petersen, and David J. Bishop, "Postexercise Cold Water Immersion Benefits Are Not Greater Than the Placebo Effect," *Medicine and Science in Sports and Exercise* 46, no. 11 (2014): 2139–47, https://doi.10.1249/MSS.0000000000000348.

19. Three Toshima Yamauchi patents relating to cryotherapy are available at https://patents.justia.com/inventor/toshima-yamauchi.

20. Darren Rovell, "Did a Mistake in New Age Ice Bath Set Back NBA Player?" CNBC, December 27, 2011, https://www.cnbc.com/id/45768144.

21. Associated Press, "Justin Gatlin Dealing with Frostbite," August 11, 2011, http://www.espn.com/olympics/trackandfield/story/_/id/6890891/justin-gatlin-arrives-world-championships-frostbite.

22. Christophe Hausswirth (n.d.), "The Effects of Whole-Body Cryotherapy Exposure in Sport: Applications for Recovery and Performance," http://skinmatrix.co.uk/image/data/cryopod/WBC_The_Science.pdf.

23. "Whole Body Cryotherapy (WBC): A 'Cool' Trend That Lacks Evidence, Poses Risks," FDA Consumer Update, July 5, 2016, accessed April 27, 2018, https://www.fda.gov/ForConsumers/ConsumerUpdates/ucm508739.htm.

24. Joseph T. Costello, Philip R. A. Baker, Geoffrey M. Minett, François Bieuzen, Ian B. Stewart, and Chris Bleakley, "Whole-Body Cryotherapy (Extreme Cold Air Exposure) for Preventing and Treating Muscle Soreness after Exercise in Adults," *Cochrane Database of Systematic Reviews* 9, no. 9 (2015): CD010789, https://doi.10.1002/14651858.CD010789.pub2.

5: Flushing the Blood

1. Antti Mero, Jaakko Tornberg, Mari Mäntykoski, and Risto Puurtinen, "Effects of Far-Infrared Sauna Bathing on Recovery from Strength and Endurance Training Sessions in Men," *SpringerPlus* 4 (2015): 321, https://doi.10.1186/s40064-015-1093-5.

2. US Food and Drug Administration official notice, August 6, 2015, "Class

2 Device Recall Portable FAR Infrared Sauna," accessed January 2018, https://www.accessdata.fda.gov/scripts/cdrh/cfdocs/cfRes/res.cfm.

3. A soigneur is a member of a cycling team's staff who provides massage and other training support to the athletes. The term comes from the French word *soigner*—to treat or tend to.

4. Jan Wilke, Robert Schleip, Werner Klingler, and Carla Stecco, "The Lumbodorsal Fascia as a Potential Source of Low Back Pain: A Narrative Review," *BioMed Research International* (2017), doi:10.1155/2017/5349620.

5. James D. Young, Alyssa-Joy Spence, and David G. Behm, "Roller Massage Decreases Spinal Excitability to the Soleus," *Journal of Applied Physiology* 124, no. 4 (April 2018), https://www.physiology.org/doi/pdf/10.1152/japplphysiol.00732.2017.

6. M. T. García-Gutiérrez, P. Guillén-Rogel, D. J. Cochrane, and P. J. Marín, "Cross Transfer Acute Effects of Foam Rolling with Vibration on Ankle Dorsiflexion Range of Motion," *Journal of Musculoskeletal and Neuronal Interactions* (2017), http://www.ismni.org/jmni/accepted/jmni_aa_GUTIERREZ.pdf.

7. Behm told me about this research in an interview. It's also detailed in an article by Cary Groner, "The Mechanistic Mysteries of Foam Rolling," *Lower Extremity Review* (October 2015), accessed May 10, 2018, http://lermagazine.com/cover_story/the-mechanistic-mysteries-of-foam -rolling.

8. Among the most promising studies on foam rolling, at least at first glance, is a 2014 study that concluded that foam rolling reduced soreness 72 hours after a hard bout of squats. The ten men who did 20 minutes of foam rolling after the workout also performed better on a vertical jump test and had better range of motion compared to the ten who didn't roll. A 2015 study by some of the same researchers found that foam rolling reduced delayed-onset muscle soreness and improved performance measures, but both of the studies were small and they used inappropriate statistical analyses, which make the conclusions suspect. Graham Z. Macdonald, Duane C. Button, Eric J. Drinkwater, and David George Behm, "Foam Rolling as a Recovery Tool after an Intense Bout of Physical Activity," *Medicine and Science in Sports and Exercise* 46, no. 1 (2014): 131–42, https://doi.10.1249/MSS.0b013e3182a123db. Meanwhile, a small 2014 study found

no performance benefit from foam rolling. Kellie C. Healey, Disa L. Hatfield, Peter Blanpied, Leah R. Dorfman, and Deborah Riebe, "The Effects of Myofascial Release with Foam Rolling on Performance," *Journal of Strength and Conditioning Research* 20, no. 1 (2014), https://doi.10.1519/JSC.0b013e3182956569.

9. Jessica Hill, Glyn Howatson, Ken van Someren, Jonathan Leeder, and Charles Pedlar, "Compression Garments and Recovery from Exercise-Induced Muscle Damage: A Meta-Analysis," *British Journal of Sports Medicine* 48, no. 18 (2014): 1340–46, https://doi.10.1136/bjsports-2013-092456.

10. Freddy Brown, Conor Gissane, and Glyn Howatson, "Compression Garments and Recovery from Exercise: A Meta-Analysis," *Sports Medicine* 47, no. 11 (2017): 2245–67, https://doi.10.1007/s40279-017-0728-9.

11. Monèm Jemni, William A. Sands, Françoise Friemel, and Paul Delamarche, "Effect of Active and Passive Recovery on Blood Lactate and Performance during Simulated Competition in High Level Gymnasts," *Canadian Journal of Applied Physiology* 28, no. 2 (2003): 240–56, https://doi.10.1139/h03-019; Egla-Irina D. Lopez, James M. Smoliga, and Gerald S. Zavorsky, "The Effect of Passive Versus Active Recovery on Power Output Over Six Repeated Wingate Sprints," *Research Quarterly for Exercise and Sport* 85, no. 4 (2014): 519–26, https://doi.10.1080/02701367.2014.961055; Gillian E. White and Greg D. Wells, "The Effect of On-Hill Active Recovery Performed Between Runs on Blood Lactate Concentration and Fatigue in Alpine Ski Racers," *Journal of Strength and Conditioning Research* 29, no. 3 (2015): 800–806, https://doi.10.1519/JSC.0000000000000677.

6: Calming the Senses

1. According to sports psychologist Göran Kenttä, "Any fool can go train more. It takes courage to rest" is an old Swedish saying.

2. *The Simpsons*, "Make Room for Lisa," season 10, episode 16, directed by Matthew Nastuk, written by Brian Scully (February 28, 1999), https://www.imdb.com/title/tt0781978/, http://www.simpsonsworld.com/video/473093187636.

3. Andrew Revkin, "John C. Lilly Dies at 86; Led Study of Communication with Dolphins," *New York Times*, October 7, 2001, http://www.nytimes

.com/2001/10/07/us/john-c-lilly-dies-at-86-led-study-of-communication
-with-dolphins.html.

4. Piritta S. Ruuska, Arto J. Hautala, Antti M. Kiviniemi, Timo H. Mäkikal-
lio, and Mikko P. Tulppo, "Self-Rated Mental Stress and Exercise Training
Response in Healthy Subjects," *Frontiers in Physiology* 3 (2012), https://
doi.10.3389/fphys.2012.00051.

5. Bryan Mann, Kirk Bryant, Brick Johnstone, Patrick Ivey, and Stephen Sayers, "The Effect of Physical and Academic Stress on Illness
and Injury in Division 1 College Football Players," *Journal of Strength
and Conditioning Research* (May 2015): 20–25, https://doi.10.1519/JSC
.0000000000001055.

6. Ryan Hall also competed, though the former elite who'd retired from overtraining wasn't running to win. And he didn't—Wardian kicked his butt.

7. Erin Strout, "Michael Wardian Wins World Marathon Challenge in
Record Time," January 29, 2017, accessed March 11, 2018, http://
www.runnersworld.com/elite-runners/michael-wardian-wins-world
-marathon-challenge-in-record-time.

8. Eliza Barclay, "Meditation Is Thriving under Trump. A Former Monk
explains why," *Vox*, July 2, 2017, accessed March 11, 2018, https://www.vox
.com/science-and-health/2017/6/19/15672864/headspace-puddicombe
-trump.

9. Matthew W. Driller and Christos K. Argus, "Flotation Restricted Environmental Stimulation Therapy and Napping on Mood State and Muscle Soreness
in Elite Athletes: A Novel Recovery Strategy?" *Performance Enhancement &
Health* 5, no. 2 (2016), https://doi.10.1016/j.peh.2016.08.002.

10. S. Bood, U. Sundequist, A. Kjellgren, T. Norlander, L. Nordström, K.
Nordenström, et al., "Eliciting the Relaxation Response with the Help
of Flotation-Rest (Restricted Environmental Stimulation Technique) in
Patients with Stress-Related Ailments," *International Journal of Stress
Management* 13, no. 2 (2006): 154.

11. Evie Stevens's new hour record was 47.980 km. Her record was broken in
September 2018 by Vittoria Bussi of Italy, who rode 48.007 km.

7: The Rest Cure

1. R. Leproult and E. Van Cauter, "Effect of 1 Week of Sleep Restriction on Testosterone Levels in Young Healthy Men," *Journal of the American Medical
Association* 305, no. 21 (2011): 2173–74, https://doi.10.1001/jama.2011.709.

2. Though it has the appearance of a children's book, *The Tortoise & The Hare & Tom Brady* is more likely meant for adults, with its wry humor and winks to Brady's earnestness. It's published by Under Armour, with illustrations by Jorge Lacera and writing by the comedy group Funny or Die, accessed March 12, 2018, http://s7d4.scene7.com/is/content/Underarmour/V7/Special%20Landers/TB12/LP/TB12_Bedtime_Story.pdf.

3. Anne-Marie Chang, Daniel Aeschbach, Jeanne F. Duffy, and Charles A. Czeisler, "Evening Use of Light-Emitting eReaders Negatively Affects Sleep, Circadian Timing, and Next-Morning Alertness," *Proceedings of the National Academy of Sciences* 112, no. 4 (2015): 1231–37, https://doi.10.1073/pnas.1418490112.

4. A. Green, M. Cohen-Zion, A. Haim, and Y. Dagan, "Evening Light Exposure to Computer Screens Disrupts Human Sleep, Biological Rhythms, and Attention Abilities," *Chronobiology International* (May 2017): 1–11, https://doi.10.1080/07420528.2017.1324878.

5. Genshiro A. Sunagawa, Kenta Sumiyama, Maki Ukai-Tadenuma, Dimitri Perrin, Hiroshi Fujishima, Hideki Ukai, Osamu Nishimura, et al., "Mammalian Reverse Genetics without Crossing Reveals Nr3a as a Short-Sleeper Gene," *Cell Reports* 14, no. 3 (2016): 662–77, https://doi.10.1016/j.celrep.2015.12.052.

6. Timothy Roehrs, Eleni Burduvali, Alicia Bonahoom, Christopher Drake, and Thomas Roth, "Ethanol and Sleep Loss: A 'Dose' Comparison of Impairing Effects," *Sleep* 26, no. 8 (2003): 981–85.

7. Aric A. Prather, Denise Janicki-Deverts, Martica H. Hall, and Sheldon Cohen, "Behaviorally Assessed Sleep and Susceptibility to the Common Cold," *Sleep* 38 (2015): 1353–59, https://doi.10.5665/sleep.4968.

8. Matthew P. Walker and Robert Stickgold, "It's Practice, with Sleep, That Makes Perfect: Implications of Sleep-Dependent Learning and Plasticity for Skill Performance," *Clinics in Sports Medicine* 24 (2005): 301–17, https://doi.10.1016/j.csm.2004.11.002.

9. Pierrick J. Amal, Thomas Lapole, Mégane Erblang, Mathias Guillard, Cyprien Bourrilhon, Damien Léger, Mounir Chennaoui, and Guillaume Y. Millet, "Sleep Extension before Sleep Loss: Effects on Performance and Neuromuscular Function," *Medicine and Science in Sports and Exercise* 48, no. 8 (2016): 1595–1603, https://doi.10.1249/MSS.0000000000000925.

10. Kelly Glazer Baron, Sabra Abbott, Nancy Jao, Natalie Manalo, and Rebecca Mullen, "Orthosomnia: Are Some Patients Taking the Quantified Self Too

Far?" *Journal of Clinical Sleep Medicine* 13, no. 2 (2017): 351–54, https://doi.10.5664/jcsm.6472.

11. Christina Draganich and Kristi Erdal, "Placebo Sleep Affects Cognitive Functioning," *Journal of Experimental Psychology: Learning, Memory, and Cognition* 40, no. 3 (2014): 857–64, https://doi.10.1037/a0035546.

12. Cheri D. Mah, Kenneth E. Mah, Eric J. Kezirian, and William C. Dement, "The Effects of Sleep Extension on the Athletic Performance of Collegiate Basketball Players," *Sleep* 34, no. 7 (2011): 943–50, https://doi.10.5665/SLEEP.1132.

13. Howard Beck, "Bowing to Body Clocks, N.B.A. Teams Sleep In," *New York Times*, December 19, 2009.

14. John Meyer, "Mikaela Shiffrin Breaks Another Record with a Win Saturday in Slalom," *Denver Post*, March 10, 2018, accessed March 11, 2018, https://www.denverpost.com/2018/03/10/mikaela-shiffrin-breaks-another-record-with-a-win-saturday-in-slalom/.

15. Gordy Megro, "Lindsey Vonn Needs a Nap, But First She Is Going to Become the Best Ski Racer of All Time," *Ski Magazine*, December 21, 2016, accessed March 11, 2018, https://www.skimag.com/ski-performance/vonn-needs-nap.

16. Tempur Sealy International, Inc., "Tempur-Pedic® Sleep Center Opens at U.S. Ski and Snowboard Association's Center Of Excellence," July 22, 2015, accessed March 11, 2018, http://www.prnewswire.com/news-releases/tempur-pedic-sleep-center-opens-at-us-ski-and-snowboard-associations-center-of-excellence-300117381.html.

17. Jonathan Abrams, "Napping on Game Day Is Prevalent Among N.B.A. Players," *New York Times*, March 6, 2011, accessed March 11, 2018, http://www.nytimes.com/2011/03/07/sports/basketball/07naps.html.

18. Roger S. Smith, Bradley Efron, Cheri D. Mah, and Atul Malhotra, "The Impact of Circadian Misalignment on Athletic Performance in Professional Football Players," *Sleep* 36, no. 2 (2013), https://doi.10.5665/sleep.3248.

8: Selling Snake Oil

1. A 2012 report published by the Centers for Disease Control and Prevention used blood and urine samples taken from a large sample of the US population to examine whether they were getting enough essential vita-

mins and nutrients. The answer was yes. The report found that nine out of ten Americans were meeting their nutritional needs. The most common exceptions were vitamin D—31 percent of non-Hispanic black Americans and 3 percent of white Americans were not getting enough— and iron, which about 8 percent of the women surveyed were deficient in. Iron deficiency is not unheard of among athletes, and women of child-bearing age are particularly prone to it and should be tested if they're experiencing unusual fatigue. But iron is rarely the nutrient being promoted in sports supplements. Centers for Disease Control and Prevention, "The Second National Report on Biochemical Indicators of Diet and Nutrition in the U.S. Population," 2012, p. 317, Table 3.1.d.1. Serum ferritin, https://www.cdc.gov/nutritionreport/.

2. Amy Eichner and Travis Tygart, "Adulterated Dietary Supplements Threaten the Health and Sporting Career of Up-and-Coming Young Athletes," *Drug Testing and Analysis* 8, no. 3–4 (2016): 304–6, https://doi.10 .1002/dta.1899.

3. In 2016, the US Federal Trade Commission settled an investigation into Herbalife Ltd. that was sparked at least in part by hedge fund manager Bill Ackman's crusade against the company. Herbalife agreed to pay $200 million and change its business practices to avoid being labeled a pyramid scheme by regulators. Diane Bartz and Michael Flaherty, "Herbalife Settles Pyramid Scheme Case with Regulator, in Blow to Pershing's Ackman," Associated Press, July 15, 2016, accessed February 21, 2018, https:// www.reuters.com/article/us-herbalife-probe-ftc/herbalife-settles-pyramid-scheme-case-with-regulator-in-blow-to-pershings-ackman-idUSKCN0ZV1F7.

4. Mina Kimes, "Drew Brees Has a Dream He'd Like to Sell You," *ESPN Magazine*, March 15, 2016, accessed December 15, 2017, http://www.espn .com/espn/feature/story/_/id/14972197/questions-surround-advocare -nutrition-empire-endorsed-saints-qb-drew-brees.

5. "Questions and Answers: Interview with Jose Antonio, Ph.D.," *Examine .com Research Digest*, no. 1 (November 2014).

6. Conor Heffernan, "Soy, Science and Selling: Bob Hoffman's Hi-Proteen Powder," Physical Culture Study, June 15, 2016, accessed March 6, 2018, https://physicalculturestudy.com/2016/06/15/soy-science-and-selling -bob-hoffmans-hi-proteen-powder/.

7. An account of Hoffman's influence in early sports nutrition and the gen-

esis of his protein products, and how he appropriated ideas from others in the field, can be found in the following paper: Daniel Hall, John D. Fair, and Frank Zane, "The Pioneers of Protein," *Better Nutrition* 8 (May–June 2004): 23–34, http://library.la84.org/SportsLibrary/IGH/IGH0803/IGH0803d.pdf.

8. Elaine Wong and Rebecca Williams, "ClinicalTrials.gov: Requirements and Implementation Strategies," Regulatory Focus, May 2012, accessed March 11, 2018, https://prsinfo.clinicaltrials.gov/publications/Wong-Williams-RAPS-Regulatory-Focus-8May2012.html.

9. Robert M. Kaplan and Veronica L. Irvin, "Likelihood of Null Effects of Large NHLBI Clinical Trials Has Increased over Time," *PLOS One* 10, no. 8 (2015), http://journals.plos.org/plosone/article?id=10.1371/journal.pone.0132382.

10. Jeffrey Beall, "What I Learned from Predatory Publishers," *Biochemia Medica* 27, no. 2 (June 2017): 273–78. doi: 10.11613/BM.2017.029.

11. I first talked with attorney Howard Jacobs while writing about my former University of Colorado cycling teammate, Tyler Hamilton. Hamilton tested positive for blood doping at the 2004 summer Olympics in Athens, where he'd won a silver medal. (He produced a second positive test following his win on Stage 8 of the Vuelta a España.) Hamilton insisted that he was innocent, and Jacobs stood by his client. Christie Aschwanden, "I Believe . . . ," *Bicycling Magazine*, November 2007, https://christieaschwanden.files.wordpress.com/2010/09/believe.pdf. In 2012, Hamilton went on *60 Minutes* and confessed to doping (and accused Lance Armstrong of doping too). Tyler, whom I once considered a friend, has never apologized or acknowledged that he lied to my face.

12. "Flash! Kicker Vencill Wins Suit against Nutrition Company, Awarded almost $600K," *Swimming World*, May 13, 2005, accessed February 15, 2018, https://www.swimmingworldmagazine.com/news/flash-kicker-vencill-wins-suit-against-nutrition-company-awarded-almost-600k/. After the verdict, the case was settled. As part of that settlement, Vencill agreed to have the judgment vacated. "It had nothing at all to do with the merits of the case, it was merely a condition of the settlement," Jacobs says. "Kicker Vencill was very happy with the terms of the settlement."

13. According to an AdvoCare press release published July 25, 2008, "Contrary to any false and misleading reports, AdvoCare products contain no ingredients banned by the United States Anti-Doping Agency (USADA) or

the World Anti-Doping Agency (WADA) which monitor Olympic and amateur sports, or by the NCAA, NFL, MLB, NBA, NHL, MLS, or NASCAR." Accessed April 27, 2018. The company sued Hardy, claiming defamation. According to Hardy's lawyer, Howard Jacobs, AdvoCare's lawsuit against her was ultimately dismissed, while her case against the company was settled. He added, "Jessica was very pleased with the settlement."

14. CAS 2009/A/1870 World Anti-Doping Agency (WADA) v. Jessica Hardy & United States Anti-Doping Agency (USADA), Arbitral Award Delivered by the Court of Arbitration for Sport. Accessed April 27, 2018, https://www.usada.org/wp-content/uploads/hardy_jessica_CAS_decision_supplement411.pdf.

15. Teri Thompson, Bill Madden, Christian Red, Michael O'Keeffe, and Nathaniel Vinton, "*Daily News* Uncovers Bizarre Plot by San Francisco Giants' Melky Cabrera to Use Fake Website and Duck Drug Suspension," *New York Daily News*, Sunday, August 19, 2012, accessed February 15, 2018, http://www.nydailynews.com/sports/baseball/exclusive-daily-news -uncovers-bizarre-plot-melky-cabrera-fake-website-duck-drug- suspension-article-1.1139623.

16. Eichner and Tygart, "Adulterated Dietary Supplements."

17. The US Anti-Doping Agency provides a library of information about supplements and publishes a current list of "high risk" products at a supplements website, accessed February 15, 2018, https://www.usada.org/substances/supplement-411/.

18. "IOC Nutritional Supplements Study Points to Need for Greater Quality Control," IOC press release, April 4, 2002, accessed February 15, 2018, https://www.olympic.org/news/ioc-nutritional-supplements-study -points-to-need-for-greater-quality-control.

19. "Health Risks of Protein Drinks: You Don't Need the Extra Protein or the Heavy Metals Our Tests Found," *Consumer Reports*, July 2010, accessed February 15, 2018, http://www.consumerreports.org/cro/2012/04/protein -drinks/index.htm.

20. Cloud and Jovanovic submitted tests of a whey protein product showing that it contained metabolites of nandrolone (a banned steroid) not listed on the label. The test results were televised nationally on Bryant Gumbel's HBO *Inside Sports* show. Mike Greenwood, Douglas S. Kalman, and Jose Antonio, *Nutritional Supplements in Sports and Exercise* (2008), https://doi.10.1007/978-1-59745-231-1.

21. Andrew I. Geller, Nadine Shehab, Nina J. Weidle, Maribeth C. Lovegrove, Beverly J. Wolpert, Babgaleh B. Timbo, Robert P. Mozersky, and Daniel S. Budnitz, "Emergency Department Visits for Adverse Events Related to Dietary Supplements," *New England Journal of Medicine* 373, no. 16 (2015): 1531–40, https://doi.10.1056/NEJMsa1504267.

22. On December 7, 2011, the US Department of Defense banned all products containing DMAA from being sold on military bases. Allied Communications Publication, ALFOODACT 036-2011 UPDATE/CORRECTION for ALFOODACT 034-2011 Dimethylamylamine (DMAA) is placed on medical hold due to possible serious adverse health effects, https://web.archive .org/web/20130302012755/http://www.troopsupport.dla.mil/subs/fso/ alfood/2011/alfo3611.pdf, accessed July 28, 2018.

23. The FDA warning about the potential dangers of Jack3d and other DMAA products was issued April 11, 2013. FDA Consumer Update, "Stimulant Potentially Dangerous to Health, FDA Warns," April 11, 2013, accessed April 27, 2018, https://www.fda.gov/ForConsumers/ConsumerUpdates/ ucm347270.htm.

24. According to an FDA statement, "On July 2, 2013, USPlabs voluntarily destroyed its DMAA-containing products located at its facility in Dallas, Texas." https://www.fda.gov/Food/DietarySupplements/ ProductsIngredients/ucm346576.htm.

25. A pending case against supplement company USPlabs charges that the company told some of its retailers and wholesalers that it used natural plant extracts in some of its products, when in fact it was using synthetic stimulants manufactured in a Chinese chemical factory, https://www.fda .gov/newsevents/newsroom/pressannouncements/ucm473099.htm. In 2013, USPlabs was also implicated in a cluster of fifty-six cases of acute liver failure or hepatitis linked to a supplement called OxyPro Elite, which the company claimed could aid muscle building and weight loss. Multiple users of the supplement required liver transplants, and one of them died.

26. Harvard researcher Peter Cohen and his colleagues have documented cases where companies that have received warning letters failed to remove the products from public sale. Peter A. Cohen, Gregory Maller, Renan DeSouza, and James Neal-Kababick, "Presence of Banned Drugs in Dietary Supplements Following FDA Recalls," *Journal of the American Medical Association* 312, no. 16 (2014): 1691–93, https://doi.10.1001/jama .2014.10308.

27. For a detailed account of how the FDA was blocked from protecting consumers from dangerous supplements, see Catherine Price, *Vitamania: How Vitamins Revolutionized the Way We Think about Food* (New York: Penguin Press, 2015).

28. The industry has multiple lobbying organizations, such as the Council for Responsible Nutrition. Longtime supplement industry supporter, retired US senator Orrin Hatch, repeatedly helped thwart efforts at regulating the industry, which has a large presence in Utah. He has family members working in the industry.

29. Anthony Roberts, "Inside the Nutritional Supplement Industry," Anthony Roberts blog, April 28, 2009, accessed May 1, 2018, https://anthonyrobertssteroidblog.wordpress.com/2009/04/28/inside-the-nutritional-supplement-industry/.

30. Bryan E. Denham, "Athlete Information Sources about Dietary Supplements: A Review of Extant Research," *International Journal of Sport Nutrition and Exercise Metabolism* 27, no. 4 (2017): 325–34, https://doi.10.1123/ijsnem.2017-0050.

9: Losing Your Zoom

1. In 2001, Ryan Hall became California state high school champion in the 1,600 m run with a state-record time of 4:02.

2. The finish did not qualify for an official American record, however, because IAAF rules disqualify point-to-point races from records to prevent runners from getting an unfair benefit from a tailwind, such as the one that had blown the day of Hall's spectacular performance, or in the case of Boston, a net downhill course. See http://running.competitor.com/2012/04/news/should-the-boston-marathon-be-record-legal_50540.

3. The New York City Marathon was canceled that year because of Hurricane Sandy.

4. Romain Meeusen, Martine Duclos, Carl Foster, Andrew Fry, Michael Gleeson, David Nieman, John Raglin, Gerard Rietjens, Jürgen Steinacker, and Axel Urhausen, "Prevention, Diagnosis, and Treatment of the Overtraining Syndrome: Joint Consensus Statement of the European College of Sport Science and the American College of Sports Medicine," *Medicine and Science in Sports and Exercise* 45, no. 1 (2013): 186–205, https://doi.10.1249/MSS.0b013e318279a10a.

5. Kevin Selby, "Ryan Hall Explains His Switch to Coach Renato Canova," *Flotrack*, December 7, 2012, accessed March 11, 2018, http://www.flotrack .org/video/666842-ryan-hall-explains-his-switch-to-coach-renato-canova.

6. S. Parker, P. Brukner, and M. Rosier, "Chronic Fatigue Syndrome and the Athlete," *Sports Medicine and Training Rehabilitation* 6 (1996): 269–78, https://www.researchgate.net/publication/232895979_Chronic_fatigue_ syndrome_and_the_athlete.

10: The Magic Metric

1. W. A. Sands, K. P. Henschen, and B. B. Schultz, "National Women's Tracking Program," *Technique*, accessed March 11, 2018, http://www .advancedstudyofgymnastics.com/uploads/3/1/9/3/31937121/1230.pdf.

2. Mike Wardian also has the world record for fastest marathon (2:38:04) run while dressed as Elvis with his win at the 2016 Rock 'n' Roll Marathon (in Las Vegas, of course).

3. The United States Department of Agriculture–Agricultural Research Service's National Nutrient Database for Standard Reference shows that raw blueberries have 2.4 grams of total dietary fiber per 100 grams (or 3.6 g per cup), whereas blackberries have 5.3 or 7.6 grams for the same amount. For blueberries, see https://ndb.nal.usda.gov/ndb/foods/ show/301068?manu=&fgcd=&ds=; for blackberries, see https://ndb.nal .usda.gov/ndb/foods/show/301063?manu=&fgcd=&ds=.

4. I accepted Quest's offer of a free trial under the condition that I couldn't promise I'd write about it or write anything positive.

5. W. S. A. Smellie, "When Is 'Abnormal' Abnormal? Dealing with the Slightly Out of Range Laboratory Result," *Journal of Clinical Pathology* 59, no. 10 (2006): 1005–7, https://doi.10.1136/jcp.2005.035048.

6. Training Peaks, "Training Stress Scores (TSS) Explained," accessed March 11, 2018, https://help.trainingpeaks.com/hc/en-us/articles/2040 71944-Training-Stress-Scores-Explained.

7. E. W. Banister and T. W. Calvert, "Planning for Future Performance: Implications for Long Term Training," *Canadian Journal of Applied Sport Sciences (Journal Canadien des Sciences Appliquees au Sport)* 5, no. 3 (1980): 170–76, http://www.ncbi.nlm.nih.gov/pubmed/6778623.

8. M. K. Drew and C. Purdam, "Time to Bin the Term 'Overuse' Injury: Is 'Training Load Error' a More Accurate Term?" *British Journal of Sports Medicine* (February 2016), https://doi.10.1136/bjsports-2015-095543.

9. Training load is a relative thing, says Tim Gabbett, an Australian sport scientist who consults with sports teams around the world. He recently published a paper arguing that it's not the training load in absolute terms that matters for injury prevention, but rather the route taken to get there. What he's found is that preseason training is crucial. "For every 10 extra sessions you complete in the preseason, you reduce your risk of injury in season by 17 percent," he says. High volumes of training can actually be protective against injury, but only if the athletes get to this level gradually, without any huge jumps in training load, Gabbett says. The acute load—the amount of work the athletes have done in the span of the week—should be in line with what they have prepared for over the previous month, Tim Gabbett, Billy Hulin, Peter Blanch, and Rod Whiteley, "High Training Workloads Alone Do Not Cause Sports Injuries: How You Get There Is the Real Issue," *British Journal of Sports Medicine* 50 (2016), doi:10.1136/bjsports-2015-095567.

10. William P. Morgan, D. R. Brown, J. S. Raglin, P. J. O'Connor, and K. A. Ellickson, "Psychological Monitoring of Overtraining and Staleness," *British Journal of Sports Medicine* 21, no. 3 (1987): 107–14, https://doi.10.1136/bjsm.21.3.107.

11. Anna E. Saw, Luana C. Main, and Paul B. Gastin, "Monitoring the Athlete Training Response: Subjective Self-Reported Measures Trump Commonly Used Objective Measures: A Systematic Review," *British Journal of Sports Medicine* (2015). https://doi.10.1136/bjsports-2015-094758.

12. Matt Dixon, *The Well-Built Triathlete* (VeloPress, 2014).

13. Inkinen's athletic feats aren't confined to triathlon. In 2014, he and his wife, Meredith Loring (a gymnast and marathoner), rowed more than 2,400 miles from San Francisco to Hawaii, unsupported. "Couple Rows across Pacific, Doesn't Divorce," read the *USA Today* headline.

14. Megan Janssen, "Breakouts, Breakdowns and Bib Offerings at the 2017 Western States 100," *Trail Runner* (June 27, 2017), accessed May 13, 2018, https://trailrunnermag.com/people/culture/breakouts-breakdowns-bib -offerings-2017-western-states-100.html.

11: Hurts So Good

1. Shona L. Halson and David T. Martin, "Lying to Win—Placebos and Sport Science," *International Journal of Sports Physiology and Performance* 8 (2013): 597–99, http://www.ncbi.nlm.nih.gov/pubmed/24194442.

2. Howard L. Fields, "The Mechanism of Placebo Analgesia," *Lancet* (September 23, 1978): 654–57, https://doi.10.1016/S0140-6736(78)92762-9.

3. David C. Nieman, Dru A. Henson, Charles L. Dumke, Kevin Oley, Steven R. McAnulty, J. Mark Davis, E. Angela Murphy, et al., "Ibuprofen Use, Endotoxemia, Inflammation, and Plasma Cytokines during Ultramarathon Competition," *Brain, Behavior, and Immunity* 20, no. 6 (2006): 578–84, https://doi.10.1016/j.bbi.2006.02.001. I first wrote about runners' unwillingness to believe Nieman's findings in 2010. Christie Aschwanden, "When It Comes to New Treatment Guidelines for Breast Cancer, Back Pain and Other Maladies, It's the Narrative Presentation That Matters," *Pacific Standard* (April 2010), accessed February 15, 2018, https://psmag.com/social-justice/convincing-the-public-to-accept-new-medical-guidelines-11422.

4. Nieman, et al., "Ibuprofen Use, Endotoxemia, Inflammation, And Plasma Cytokines during Ultramarathon Competition"; Steven R. McAnulty, John T. Owens, Lisa S. McAnulty, David C. Nieman, Jason D. Morrow, Charles L. Dumke, and Ginger L. Milne, "Ibuprofen Use during Extreme Exercise," *Medicine and Science in Sports and Exercise* 39, no. 7 (2007): 1075–79, https://doi.10.1249/mss.0b13e31804a8611.

5. Claire Baxter, Lars R. McNaughton, Andy Sparks, Lynda Norton, and David Bentley, "Impact of Stretching on the Performance and Injury Risk of Long-Distance Runners," *Research in Sports Medicine* 25, no. 1 (2017): 78–90, https://doi.10.1080/15438627.2016.1258640.

6. Many people who learn that they were in a placebo or sham treatment group still insist that they were helped by whatever they got, says Tor Wager, a placebo researcher at the University of Colorado. "There's a lot of room for people to say, well, I think it works for me."

7. Claudia Carvalho, Joaquim Machado Caetano, Lidia Cunha, Paula Rebouta, Ted J. Kaptchuk, and Irving Kirsch, "Open-Label Placebo Treatment in Chronic Low Back Pain." *Pain* (October 2016): 1, https://doi.10.1097/j.pain.0000000000000700.

8   J. P. Rose, A. L. Geers, H. M. Rasinski, and S. L. Fowler, "Choice and Placebo Expectation Effects in the Context of Pain Analgesia," *Journal of Behavioral Medicine* 35, no. 4 (August 2012): 462–70, doi: 10.1007/s10865-011-9374-0.

9. Alex Stone, "Why Waiting Is Torture," *New York Times*, August 19, 2012,

accessed March 11, 2018, http://www.nytimes.com/2012/08/19/opinion/sunday/why-waiting-in-line-is-torture.html.

10. Martin D. Hoffman, Natalie Badowski, Joseph Chin, and Kristin J. Stuempfle, "A Randomized Controlled Trial of Massage and Pneumatic Compression for Ultramarathon Recovery," *Journal of Orthopaedic and Sports Physical Therapy* 46, no. 5 (2016): 1–26, https://doi.10.2519/jospt.2016.6455.

11. Jonas Bloch Thorlund, "Deconstructing a Popular Myth: Why Knee Arthroscopy Is No Better Than Placebo Surgery for Degenerative Meniscal Tears," *British Journal of Sports Medicine* (2017), https://doi.10.1136/bjsports-2017-097877.

12. Steve Magness, "When Doing Nothing Is Better Than Doing Something," *Science of Running* (April 2016), accessed March 11, 2018, http://www.scienceofrunning.com/2016/04/when-doing-nothing-is-better-than-doing.html.

13. James Hamblin, "Please, Michael Phelps, Stop Cupping," *The Atlantic*, August 9, 2016, accessed March 11, 2018, https://www.theatlantic.com/health/archive/2016/08/phelps-cupsanity/495026/.

14. Taylor Phinney's parents are Connie Carpenter-Phinney and Davis Phinney. Connie competed as a speed skater in the 1972 Winter Olympics at age 14. At the 1984 Summer Olympics in Los Angeles, she competed as a cyclist and won a gold medal in the road race. Davis spent more than a decade as a professional cyclist. He won two stages of the Tour de France in the 1980s.

15. Sam Alipour, "Will You Still Medal in the Morning?" *ESPN the Magazine*, July 8, 2012, accessed February 15, 2018, http://www.espn.com/olympics/summer/2012/story/_/id/8133052/athletes-spill-details-dirty-secrets-olympic-village-espn-magazine.

Conclusion

1. Thomas M. Doering, David G. Jenkins, Peter R. Reaburn, Nattai R. Borges, Erik Hohmann, and Stuart M. Phillips, "Lower Integrated Muscle Protein Synthesis in Masters Compared with Younger Athletes," *Medicine and Science in Sports and Exercise* 48, no. 8 (2016): 1613–18, https://doi.10.1249/MSS.0000000000000935.

INDEX